Durability and Sustainability of Concrete

Nausherwan Hasan

Durability and Sustainability of Concrete

Case Studies for Concrete exposures

Nausherwan Hasan
AECOM
New York, NY, USA

ISBN 978-3-030-51575-1 ISBN 978-3-030-51573-7 (eBook)
https://doi.org/10.1007/978-3-030-51573-7

This Springer imprint is published by the registered company Springer Nature Switzerland AG
The registered company address is: Gewerbestrasse 11, 6330 Cham, Switzerland

Conversion Factors

SI* (modern metric) conversion factors									
Approximate conversions to SI units					Approximate conversions from SI units				
Symbol	When you know	Multiply by	To find	Symbol	Symbol	When you know	Multiply by	To find	Symbol
Length					**Length**				
in	Inches	25.4	Millimeters	mm	mm	Millimeters	0.039	Inches	in
ft	Feet	0.305	Meters	m	m	Meters	3.28	Feet	ft
yd	Yards	0.914	Meters	m	m	Meters	1.09	Yards	yd
mi	Miles	1.61	Kilometers	km	km	Kilometers	0.621	Miles	mi
Area					**Area**				
in2	Square inches	645.2	Millimeters squared	mm2	mm2	Millimeters squared	0.0016	Square inches	in2
ft2	Square feet	0.093	Meters squared	m2	m2	Meters squared	10.764	Square feet	ft2
yd2	Square yards	0.836	Meters squared	m2	m2	Meters squared	1.196	Square yards	yd2
ac	Acres	0.405	Hectares	ha	ha	Hectares	2.47	Acres	ac
mi2	Square miles	2.59	Kilometers squared	km2	km2	Kilometers squared	0.386	Square miles	mi2
Volume					**Volume**				
fl oz	Fluid ounces	29.57	Milliliters	ml	ml	Milliliters	0.034	Fluid ounces	floz
gal	Gallons	3.785	Liters	L	L	Liters	0.264	Gallons	gal
ft3	Cubic feet	0.028	Meters cubed	m3	m3	Meters cubed	35.315	Cubic feet	ft3
yd3	Cubic yards	0.765	Meters cubed	m3	m3	Meters cubed	1.308	Cubic yards	yd3
Note: Volumes greater than 1000 L shall be shown in m3.									
Mass					**Mass**				
oz	Ounces	28.35	Grams	g	g	Grams	0.035	Ounces	oz
lb	Pounds	0.454	Kilograms	kg	kg	Kilograms	2.205	Pounds	lb
T	Short tons (2000 lb)	0.907	Megagrams	Mg	Mg	Mega grams	1.102	Short tons (2000 lb)	T
Temperature (exact)					**Temperature (exact)**				
°F	Fahrenheit	(F-32)/1.8	Celsius	°C	°C	Celsius	1.8C + 32	Fahrenheit	°F

*SI is the symbol for the International System of Measurement.

Contents

About the Author

Nausherwan Hasan is a registered professional engineer with 40 years plus experience in civil design and construction management of nuclear, hydroelectric, water resources and infrastructure projects. His responsibilities included preparation of construction specifications for concrete, support during construction, design and control of concrete mixtures for various regions in the USA and overseas, rehabilitation of concrete dams, spillways, and canal structures, surveillance, instrumentation, and thermal monitoring, preparation and review of quality assurance/quality control procedures, contact grouting of steel and concrete lined tunnels, inspection and trouble-shooting reports, expert witness on concrete technology, condition survey and inspection of dams and spillways. He has published over a dozen technical papers and articles on concrete technology and participated in international conferences on concrete durability. Currently, he is a member of American Concrete Institute and ASTM technical committees, and a Fellow of American Society of Civil Engineers.

Abbreviations and Acronyms

ACI	American Concrete Institute
AAR	Alkali-Aggregate Reactivity
AASHTO	American Association State Highway Transportation Officials
ACR	Alkali-Carbonate Reactivity
AEA	Air-Entraining Admixture
ANSI	American National Standards Institute
ASR	Alkali-Silica Reactivity
ASME	American Society of Mechanical Engineers
ASTM	American Society of Testing and Materials
AWA	Anti-Washout Admixture
BASF	BASF Corporation (Admixtures Manufacturer)
CA	Coarse Aggregate
CF	Coarseness Factor
CGDP	Commercial Grade Dedication Plan
CLSM	Controlled Low Strength Material
CRD-C	Corps of Engineers Standard Designation
CRR	Corrosion Resistant Steel
COE	Corps of Engineers
CPMB	Concrete Plant Manufacturers Bureau
DOT	Department of Transportation
EADP	Emergency Access Drift Plug
ECR	Epoxy Coated Reinforcement
EPA	Environmental Protection Agency
FA	Fine Aggregate
FHWA	Federal Highway Administration
FM	Fineness Modulus
GGBFS	Ground Granulated Blast Furnace Slag, or Slag Cement
HPC	High Performance Concrete
HRWRA	High Range Water-Reducing Admixture
QA	Quality Assurance
QC	Quality Control

NDT	Non-Destructive Testing
NEF	National Enrichment Facility
NRMCA	National Ready Mixed Concrete Association
NST	Nitrogen Sidestream Treatment
NMSA	Nominal Maximum Size Aggregate
NQA	Nuclear Quality Assurance
NYCDEP	New York City Department of Environmental Protection
ODOT	Oregon Department of Transportation
OSHA	Occupational Safety and Health Administration
PCA	Portland Cement Association
PPS	Prefabricated Powerplant structure
PRA	Permeability Reducing Admixture
RWP	Round-out West-Branch
SCC	Self-Consolidating Concrete
SCM	Supplementary Cementitious Material
SG	Specific Gravity
SRA	Shrinkage Reducing Admixture
SRWTP	Sacramento Regional Waste Treatment Plant
SSD	Surface Saturated Dry
TMMB	Truck Mixer Manufacturers Bureau
TPC	Thermal Control Plan
TTF	Tertiary Treatment Facility
USBR	United States Bureau of Reclamation
W/CM	Water-Cementitious Ratio

American Association of State Highway Official (AASHTO)

AASHTO T26	Standard Test Method for Quality of Water
AASHTO T 334	Standard Test Method for Estimating Crack Tendency of Concrete

American Concrete Institute (ACI)

ACI 117	Standard Specifications for Tolerances for Concrete Construction and Materials
ACI 201.2R	Guide to Durable Concrete
ACI 207.1R	Mass Concrete.
ACI 210R	Erosion of Concrete in Hydraulics Structures
ACI 211.1	Recommended Practice for Selecting Proportions for Normal Weight Concrete
ACI 212.3R	Report on Chemical admixtures for Concrete

ACI 214	Recommended Practice for Evaluation of Strength Test Results of Concrete
ACI 214.3R	implified Version of the Recommended Practice for Evaluation of Strength Test Results of Concrete
ACI 221.4	Report on Alkali-Aggregate reactivity
ACI 222R	Protection of Metals in Concrete against Corrosion
ACI 222.3R	Guide to Construction Practices to Mitigate Corrosion of Reinforcement in Concrete structures
ACI 223R	Guide to the Use of Shrinkage-Compensating Concrete
ACI 229R	Report on Controlled Low-Strength Materials (CLSM)
ACI 233R	Guide to the Use of Slag cement
ACI 224R	Control of Cracking in Concrete
ACI 237	Self-Consolidating Concrete
ACI 301	Specifications for Structural Concrete for Buildings.
ACI 304R	Guide for Measuring, Mixing, Transporting, and Placing Concrete
ACI 305R	Guide to Hot Weather Concreting
ACI 306R	Guide to Cold Weather Concreting
ACI 308R	Guide to External Curing of Concrete
ACI 309R	Guide for Consolidation of Concrete
ACI 311.4	Guide for Concrete Inspection
ACI 318	Building Code Requirements for Reinforced Concrete
ACI 347	Guide for Formwork for Concrete
ACI 349 Code	Requirements for Nuclear Safety Related Concrete Structures and Commentary
ACI 350 Code	Requirements for Environmental Engineering Concrete Structures and Commentary
ACI 359	Code for Concrete Containment
ACI 359 NQA.1	Quality Assurance Requirements for Nuclear Facility Applications
ACI 365.1R	Service Life Prediction
ACI 546R	Guide to Concrete Repair
ACI SP 2	ACI Manual of Concrete Inspection

American Society for Testing and Materials (ASTM)

ASTM C 33/C 33M	Standard Specification for Concrete Aggregates
ASTM C 39/C39M	Standard Test Method for Compressive Strength of Cylindrical Concrete Specimens
ASTM C 40/C40M	Organic Impurities in Fine Aggregates for Concrete
ASTM C 87/C87M	Effect of Organic Impurities in Fine aggregate on Strength of Mortar

ASTM C 88/C88M	Standard Test Method for Soundness of Aggregates by Use of Sodium Sulfate or Magnesium Sulfate
ASTM C 117/C117M	Standard Test Method for Materials Finer than 75-um (No. 200) Sieve in Mineral Aggregates by Washing
ASTM C125/C125M	Standard Terminology Relating to Concrete
ASTM C 127/C127M	Standard Test Method for Density, Relative Density (Specific Gravity), and Absorption of Coarse Aggregate
ASTM C 128/C128M	Standard Test Method for Density, Relative Density (Specific Gravity), and Absorption of Fine Aggregate
ASTM C 131/C131M	Standard Test Method for Resistance to Degradation of Small-Size Coarse Aggregate by Abrasion and Impact in the Los Angeles Machine
ASTM C 136/C136M	Standard Test Method for Sieve Analysis of Fine and Coarse Aggregates
ASTM C 138/C 138M	Standard Test Method for Density (Unit Weight), Yield, and Air Content (Gravimetric) of Concrete
ASTM C 142/C142M	Standard Test Method for Clay Lumps and Friable Particles in Aggregates
ASTM C 143/C 143M	Standard Test Method for Slump of Hydraulic-Cement Concrete
ASTM C 150/C 150M	Standard Specification for Portland Cement
ASTM C157/C157M	Standard Test Method for Length Change of Hardened Concrete
ASTM C 172/C172M	Standard Practice for Sampling Freshly Mixed Concrete
ASTM C 192/C192M	Standard Practice for Making and Curing Concrete Test Specimens in the Laboratory
ASTM C 173/C173M	Standard Test Method for Air Content of Freshly Mixed Concrete by the Volumetric Method
ASTM C186/C186M	Standard Test Method for Heat of Hydration of Hydraulic Cement (Withdrawn 2019)
ASTM C 231/C231M	Standard Test Method for Air Content of Freshly Mixed Concrete by the Pressure Method
ASTM C 232/C232M	Standard Test Method for Bleeding of Air Content of Freshly Mixed Concrete
ASTM C 260/C260M	Standard Specification for Air-Entraining Admixtures for Concrete
ASTM C 295/C295M	Standard Guide for Petrographic Examination of Aggregates for Concrete
ASTM C311/C311M	Sampling and Testing Fly Ash or Natural Pozzolans for Use as a Mineral Admixture in Portland-Cement Concrete
ASTM C 403/C403M	Standard Test Method for Time of Setting Concrete Mixtures by Penetration Resistance
ASTM C418/C418M	Standard Test Method for Abrasion Resistance of Concrete by Sandblasting

ASTM C441/C441M Standard Test Method for Effectiveness of Pozzolans or ground Blast-Furnace Slag in Preventing Excessive Expansion Due to the Alkali-Silica Reaction
ASTM C452/C452M Standard Test Method for Effectiveness of Pozzolans or ground Blast-Furnace Slag in Preventing Excessive Expansion Due to the Alkali-Silica Reaction
ASTM C457/C457M Standard Test Method for Potential Expansion of Portland Cement Mortars Exposed to Sulfate
ASTM C 494/C 494M Standard Specification for Chemical Admixtures for Concrete
ASTM C 535/C535M Standard Test Method for Resistance to Degradation of Large-Size Coarse Aggregate by Abrasion and Impact in the Los Angeles Machine
ASTM C566/C566M Standard Test Method for Total Evaporable Moisture Content of Aggregates by Drying
ASTM C586/C586M Standard Test Method for Potential Alkali Reactivity of Carbonate Rocks for Concrete Aggregates (rock-Cylinder Method)
ASTM C 595/C 595M Standard Specification for Blended Hydraulic Cement
ASTM C 618/C 618M Standard Specification for Coal Fly Ash and Raw or Calcined Natural Pozzolan for Use in Concrete
ASTM C642/C 642M Standard Test Method for Density, Absorption, and Voids in Hardened Concrete
ASTM C666/C666M Standard Test Method of Resistance of Concrete to Rapid Freezing and Thawing
ASTM C805/C805M Standard Test Method for Rebound Number of Hardened Concrete
ASTM C845/C845M Standard Specification for Expansive Hydraulic Cement
ASTM C856/C856M Petrographic Examination of Hardened Concrete
ASTM C 989/C 989 M Standard Specification for Ground Granulated Blast-Furnace Slag for Use in Concrete and Mortars
ASTM C 944/C 944M Standard Test Method for Abrasion Resistant of Concrete or Mortar Surfaces by the Rotating -Cutter Method Exposed to Sulfate Solution
ASTM C 1012/C 1012M Standard Test Method for Length Change of Hydraulic-Cement Mortars Exposed to a Sulfate Solution
ASTM C 1017/C 1017M Standard Specification for Chemical Admixtures for Use in Producing Flowing Concrete
ASTM C 1064/C 1074M Standard Test Method for Temperature of Freshly Mixed Concrete
ASTM C1074/C1074M Standard Practice for Estimating Concrete Strength by the Maturity Method
ASTM C1077/C 1077 M Standard Practice for Agencies Testing Concrete and Concrete aggregates for Use in Construction and Criteria for Testing Agency Evaluation

ASTM C1105/C1105M5	Standard Test Method for Determination of Length Change of Concrete Due to Alkali-Carbonate Reaction
ASTM C1157/C 1157M	Standard Performance Specification for Hydraulic Cement
ASTM C1202/C1202M	Standard Test Method for Electrical Indication of Concrete's Ability to Resist Chloride Ion Penetration
ASTM C1240/C1240M	Standard Specification for Silica Fume Used in Cementitious Mixtures
ASTM C 1260/C 1260M	Standard Test Method for Potential Alkali-Silica Reactivity of Aggregates (Mortar-Bar Method)
ASTM C1293/C1293M	Standard Test Method for Potential Alkali-Silica Reactivity of Aggregates (Mortar-Bar Method)
ASTM C1315/C1315M	Standard Specification for Liquid Membrane-Forming Compounds Having Special Properties for Curing and Sealing Concrete
ASTM C1556 /C1556M	Standard Test Method for Determining the Apparent Chloride Diffusion Coefficient of Cementitious Mixtures by Bulk Diffusion
ASTM C 1567/C 1567M	Standard Test Method for Potential Alkali-Silica Reactivity of Combinations of Cementitious Materials and Aggregate (Accelerated Mortar Bar Method)
ASTM C1581/C1582M	Standard Test Method for Determining Age at Cracking and Induced Tensile Stress Characteristics of Mortar and Concrete under Restrained Shrinkage
ASTM C1582/C1582M	Standard Specification for Admixtures to inhibit Chloride-Induced Corrosion of Reinforcing Steel in Concrete
ASTM C1602/C1602M	Standard Specification for Mixing Water Used in the Production of Hydraulic Cement Concrete
ASTM C 1611/C 1611M	Standard Test Method for Slump of Hydraulic-Cement Concrete
ASTM C1792/C1792M	Standard Test Method for Measurement of Mass Loss Versus Time for One-Dimensional Drying of Concrete
ASTM A1035/A1035M	Standard Specification for Deformed and Plain Low-carbon, Chromium, Steel Bars for Concrete reinforcement
ASTM D512/D512M	Standard Test Method for Chloride Ion in Concrete
ASTMD516/C516M	Standard Test Method for Sulfate Ion in Concrete
ASTM D1751/C1751M	Standard Specification for Preformed Expansion Joint Fillers for Concrete Paving and Structural Construction (Non-extruding and Resilient Bituminous Types
ASTM D 4791/C4791M	Flat Particles, Elongated Particles, or Flat and Elongated Particles in Coarse Aggregate
ASTM D4832/D4832M	Standard Test Method for Preparation and Testing of Controlled Low-Strength Materials (CLSM)

ASTM D 5084/C5084M Standard Test Method for Measurement of Hydraulic Conductivity of saturated Porous Materials
ASTM E 329/E 329 M Standard Specification for Agencies Engaged in the Testing and/or Inspection of Materials Used in Construction

U.S. Army Corps of Engineers (USACE)

COE CRD-C61 Test Method for Determining the Resistance of Freshly Mixed Concrete to Washing Out in Water
COE CRD-C120 Test Method for Flat and Elongated Particles in Fine Aggregate
COE CRD-C143 Specification for Meters for Automatic Indication of Moisture in Fine aggregate
COE CRD-C661 Specification for Anti-washout Admixture for Concrete
COE CRD-C662 Determining the Potential Alkali-Silica Reactivity of Combinations of Cementitious Materials, Lithium Nitrate Admixture and Aggregate (accelerated Mortar-Bar Method)
CPMB Concrete Plant Manufacturers Bureau
NRMCA National Ready Mixed Concrete Association
PCA Portland Cement Association
TMMB Truck Miser Manufacturers Bureau

Introduction

The durability of concrete structures is a well-recognized phenomenon by the presence and continued service of many structures still standing well beyond their original design life. While the modern cement was invented by Portland in 1824, the roots of this material date back to 3000 BC, when the ancient pyramids were built using gypsum and lime to make mortar for the stones. The ancient Romans used a material similar to the modern cement to build many of their architectural marvels, such as the Colosseum in Rome, more than 15 centuries ago.

The concrete industry in the USA has evolved over time. The concrete construction in the USA dates back to1903, with the first high rise building, built in Cincinnati, Ohio, in 1903.The use of ready mixed concrete mixed at a central plant occurred in 1913, and delivered by truck to the job site; this was the beginning of the concrete revolution. One of the first concrete gravity dam was the Foote dam on Au Sable River, completed in 1918.

The concrete industry is the beneficiary of both public and private resources that provided the stimulus for advances in concrete technology. Such efforts were originally spear-headed by United States Bureau of Reclamation (USBR) and Portland Cement Association (PCA), followed by the American Concrete Institute (ACI), and supported by research at various universities, state DOT's, and, in particular, US Army Corps of Engineers Research Laboratory at Vicksburg, MS. American Society of Testing and Materials (ASTM) provides the concrete industry with testing and standards for concrete.

Established in 1902, USBR is best known for the dams, power plants, and canals it constructed in the 17 western states. These water projects led to homesteading and promoted the economic development of the Wild West. Reclamation constructed more than 600 dams and reservoirs. The design and construction of Hoover Dam (1936) on the Colorado River and Grand Coulee Dam (1942) on the Columbia River led to the development of largescale improvements in concrete technology dealing with materials for concrete, mixing, placing, and cooling requirements for concrete.

The development of the cement industry in the late 1920s was dictated by the need to understand chemical processes of hydration in order to reduce cracking due to heat generated from mass concrete, deterioration from sulfate attack and alkali-aggregate reaction, and freezing and thawing environments.

USBR has accumulated database of concrete and monitoring programs, which are useful for comparing trends for various deterioration mechanisms.

Concrete resists weathering, chemical attack, and abrasion while maintaining its desired engineering properties over time. However, well-documented data of concrete operational experience and longevity is scarce. Such data is needed for use by professional engineers and the concrete industry as a basis for concrete life–extension evaluation. It is in this context, that the author felt the need to compile some historical concrete data on concrete construction in the USA to serve as a basis for future reference and to address the durability issues.

The author has been associated with the design and construction of concrete structures relating to nuclear power plants, hydroelectric power stations, and infrastructure projects for well over 40 years. This book focuses on major concrete degradation mechanisms affecting concrete and discusses the mitigation measures for durability. It identifies the appropriate cement, reinforced with chemical admixtures, and suitable fine and coarse aggregate combinations to mitigate alkali-aggregate reaction (AAR), drying shrinkage and cracking, sulfate attack, and freezing and thawing deterioration. The use of supplementary cementitious materials (SCM), including granulated ground blast-furnace slag (GGBFS), fly ash, and silica fume, which are industrial by-products in concrete, is an ecological benefit which mitigates waste disposal. It also reduces the carbon emissions due to reduction of cement in concrete.

The self-consolidating concrete (SCC) innovation is relatively new and has become successful for placing concrete in congested reinforcing locations and for underwater applications, with minimal labor and consolidation. SCC became viable by the advances in admixture technology in the 1980s, including high range water reducing and viscosity modifying admixtures.

Case studies are included from various regions of the USA to provide greater insight for selecting appropriate concrete mixtures for each specific environmental condition. These studies identify specific requirements for placing, finishing, and curing of concrete. Quality assurance and quality control procedures are also included to assure that concrete meets the design and serviceability objectives for a specific project.

According to a recent UN study, the urban population has increased to 50% for the first time in history, and it is forecast to reach 66% by 2050, the changes being driven from newly industrialized countries. The trend will increase demands of urban housing and development. The demand on concrete production, especially the cement, is on the rise. For preservation of natural resources, there is a need and demand to enhance durability and sustainability of concrete while also mitigating environmental impact. It is estimated that cement production contributes perhaps as much as 5% of man-made carbon emissions during its manufacture. Companies like CarbonCure are developing technologies for concrete manufacturing that will absorb carbon dioxide and make it stronger and greener concrete.

In the 1970s, a service life of 40–50 years was considered adequate as design basis for a nuclear power plant concrete structure, consistent with the service life of the other safety components (equipment, instrumentation, etc.). With the rising repair and maintenance costs of infrastructure, there is a need for high performance concrete and specific durability requirements for new concrete structures. To meet this challenge, a quantitative approach for durability/service life has been developed by the concrete industry.

The quantitative approach, based on finite-element modeling, is a great tool in establishing and predicting service life of concrete. The durability modeling combines materials engineering and computer simulations to optimize the reinforced concrete system in achieving a service life goal. Variables in such a model include concrete mixture proportions, design requirements, in-service degradation processes, and environmental exposure conditions. A typical scope of work for achieving a 100-year service life, which highlights concrete durability matrix requirements and a laboratory test program for achieving the desired results has been addressed with Input from Richard Cantin of Simco Technologies.

Experimental studies are underway to reduce the carbon footprint by minimizing cement in concrete, while enhancing its durability with use of SCM and fillers like limestone fines, resulting in lower heat of hydration and improved permeability of concrete. The innovation in concrete technology and a performance-based approach to designing concrete structures will further provide sustainable solutions for the concrete industry.

The author is indebted to many organizations and individuals for permission to utilize published materials and photographs. Special acknowledgment is due to ACI, ASTM, PCA, and USBR for their contribution to concrete technology. I am particularly thankful to my mentors, Andrew A Ferlito, Allan Wern, and (late) Murray Weber at Ebasco services Inc., a legacy firm of AECOM, for their encouragement and support. In particular, I am thankful to Joseph Ehasz, Enver Odar, Edward O′ Connor, and Rashil Levent for reviewing parts of the manuscript. Lastly, I am grateful to Allen Hulshizer for reviewing the entire manuscript and for his valuable suggestions.

New York City
April 2020

Chapter 1
Concrete Materials

1.1 Cementitious Materials

Cementitious material consists of Portland cement, fly ash, slag cement, and silica fume as applicable.

Portland Cement It is the basic ingredient of concrete, and a closely controlled chemical combination of calcium, silicon, aluminum, iron, and small amounts of other ingredients to which gypsum is added in the final grinding process to regulate the setting time of the concrete. Lime and silica make up about 85% of the mass, as shown in Table 1.1.

Other compounds are magnesium oxide, MgO, calcium oxide, lime, CaO, and equivalent alkalis (Na_2O + 0.658 K_2O) limited to about 1–2%.

Hydration of Cement: In the presence of water, the cement compounds chemically combined with water (hydrate) to form new compounds that are the infrastructure of the hardened cement paste in concrete. The hydration is described by the rate and the heat of hydration.

Heat of Hydration Heat of hydration is the heat generated when cement and water react. The amount of heat generated is dependent chiefly upon the chemical composition of the cement, with C_3A and C_3S being the compounds primarily responsible for high heat evolution.

Characteristics of hydration of the cement compounds are shown in Figs. 1.1 and 1.2.

C_3S: Hydrates and hardens rapidly and is largely responsible for initial set and early strength.

Early strength of cement is higher with increased percentages of C_3S.

C_2S: Hydrates and hardens slowly. Contributes largely to strength increase at ages beyond 1 week.

N. Hasan, *Durability and Sustainability of Concrete*,
https://doi.org/10.1007/978-3-030-51573-7_1

Table 1.1 Compounds of Portland cement

Notation	Description	Formula	Weight, %
C_3S	Tricalcium silicate	$3CaO \cdot SiO_2$	55
C_2S	Dicalcium silicate	$2CaO \cdot SiO_2$	18
C_3A	Tricalcium aluminate	$3CaO \cdot Al_2O_3$	10
C_4AF	Tetra-calcium alumino ferrite	$4CaO \cdot Al_2O_3$ Fe_2O_3	8
CSH_2	Gypsum	$CaSO4 \cdot 2H_2O$	6

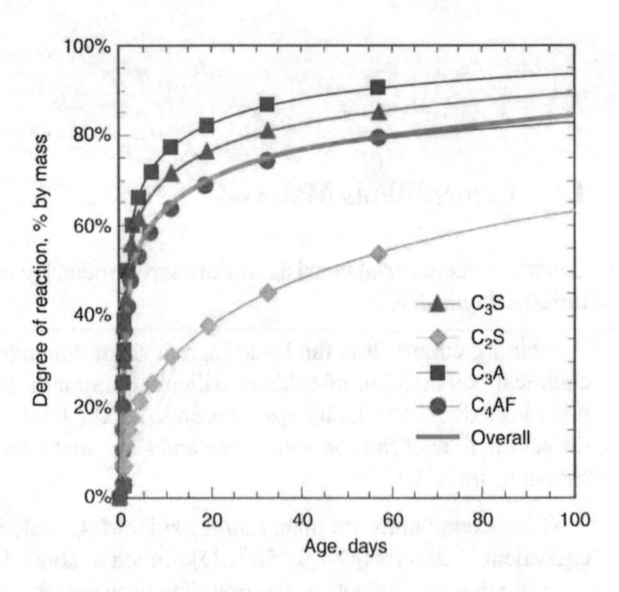

Fig. 1.1 Cement major compounds and degree of hydration with time

C_3A: Liberates a large amount of heat during the first few days of hydration and hardening. Also contributes slightly to early strength development.

Gypsum: Added to the cement slows down the hydration rate of C_3A.

Cements with low percentages of C_3A are especially resistant to soils and waters containing sulfates.

C_4AF: Does not play any significant role on hydration.

1.2 Types of Portland Cement

Portland cement should conform to ASTM C150 Type I, II, III, IV, or V.

ASTM C150 covers specification for Portland cements. Originally, this specification covered five basic types of cement. Over the years, however, the types of cement offered by industry have expanded to cover ten types of Portland cement. The five basic classifications are:

Type I: For general use

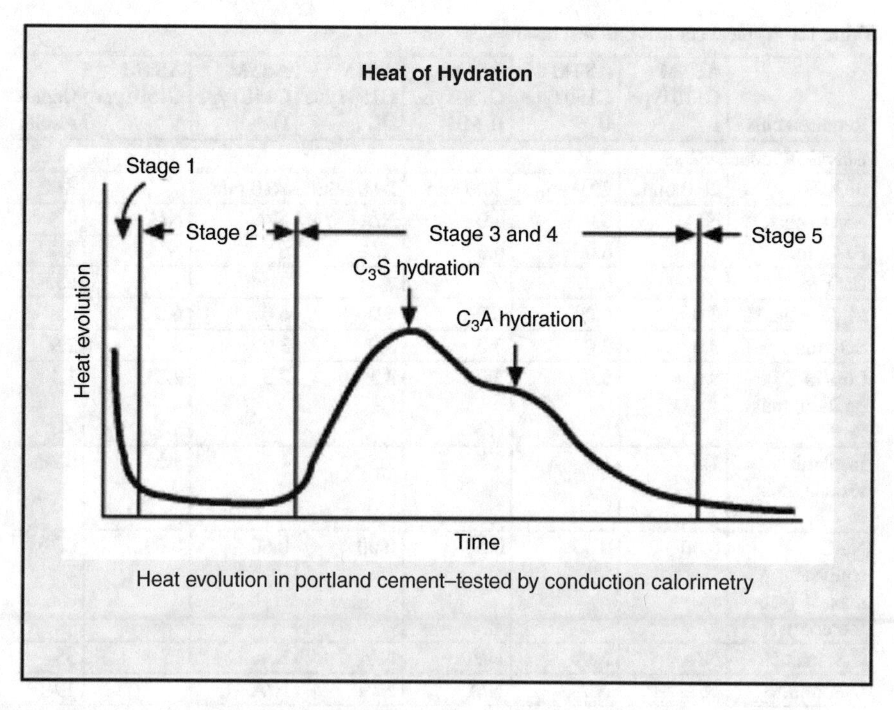

Fig. 1.2 Cement compounds C_3S and C_3A hydration stages

Type II: For general use, where moderate heat of hydration is desired
Type III: For use when high early strength is desired
Type IV: For use when a low heat of hydration is desired
Type V: For use when high sulfate resistance is desired

Other variations of types of Portland cement offered by the industry are IA, IIA, and IIIA, where air entrainment is desired. For specific moderate heat of hydration and moderate sulfate resistance, Types II(MH) and II(MH)IA have been added to ASTM C150.

Table 1.2 lists the specific chemical composition and physical requirements of the cement types. Some cements are designated with a combination Type I/II, indicating that the cement meets the both Type I and Type II cement.

Blended hydraulic cements conform to ASTM C595. Figure 1.3 shows a typical mill test certification report for the Lehigh blended cement.

For planning purposes, the local availability of the desired cement type should be verified prior to the design process. For a particular project, the type of the Portland cement should be carefully selected, based on site conditions and application. When ordering cement, the engineer should specify the type of cement to be used, including optional composition requirements such as tricalcium aluminate (C_3A), equivalent alkalis (Na_2O + $0.658K_2O$), heat of hydration, and sulfate resistance in accordance with Tables 1.2, 1.3, and 1.4 of ASTM C150.

Table 1.2 Typical cement mill test results

Requirements	ASTM C150Type I	ASTM C150Type II	ASTM C150Type II(MH)	ASTM C150Type III	ASTM C150Type IV	ASTM C150Type V	Cemex results
Chemical requirements							
SiO_2, %	20.0 min.	20.0 min.	20.0 min.	20.0 min.	20.0 min.		21.2
Al_2O_3, max. %	N/A	6.0	6.0	N/A	N/A	N/A	4.5
Fe_2O, max. %		6.0	6.0	N/A	6.5	N/A	4.4
CaO, %							63.8
MgO, max. %	6.0	6.0	6.0	6.0	6.0	6.0	0.8
SO_3, max. %	3.0	3.0	3.0	3.0	3.0	3	2.9
Loss of ignition, max. %	3.0	3.0	3.0	3.5	2.3	2.3	1.1
Insoluble residue, max. %	1.5	1.5	1.5	1.5	1.5	1.5	0.28
Na_2O equivalent, max. % (see footnote)	0.60	0.60	0.60	0.60	0.60	0.60	0.39
C_3S, max.%	N/A	N/A	N/A	N/A	N/A		55
C_2S, min.%	N/A	N/A	N/A	N/A	N/A		19
C_3A, max. %		8.0	8.0	15.0	7.0	5.0	5
C_4AF, %	N/A	N/A	N/A	N/A	N/A		13
$C_{4AF} + 2C_3A$, max. %	N/A	N/A	N/A	N/A	N/A	25	22
Physical requirements							
Blaine fineness, m^2/kg (air permeability)	260 min	260 min	260 min		260 min 430 max	260 min	382
Minus 325 mesh							96.7
Vicat set time							
Initial, minutes	45 min	45 min	45 min	45 min	45 min	45 min	89
Final, minutes	375 max.	375 max.	375 max.	375 min	375 max.	375 max.	182
False set, min. %	50	50	50	50	50	50	81
Air content, max. %	12	12	12	12	12	12	6.0
Autoclave expansion, max. %	0.80	0.80	0.80	0.80	0.80	0.80	0.003
Compressive strength, psi							
1 day				1740			2120
3 days	1740	1450	1450	3480	N/A	1160	3810

(continued)

Table 1.2 (continued)

Requirements	ASTM C150Type I	ASTM C150Type II	ASTM C150Type II(MH)	ASTM C150Type III	ASTM C150Type IV	ASTM C150Type V	Cemex results
7 days	2760	2470	2470		1020	2180	5020
28 days					2470	3050	6640

Note: ASTM C150-19a has eliminated the requirements of alkalis in cement, which was historically designed as "low-alkali cements" and recommended for use with aggregate sources susceptible to alkali-silica reaction (ASR). However, low-alkali cement alone may not be effective for the ASR mitigation. See Chap. 3 for additional information

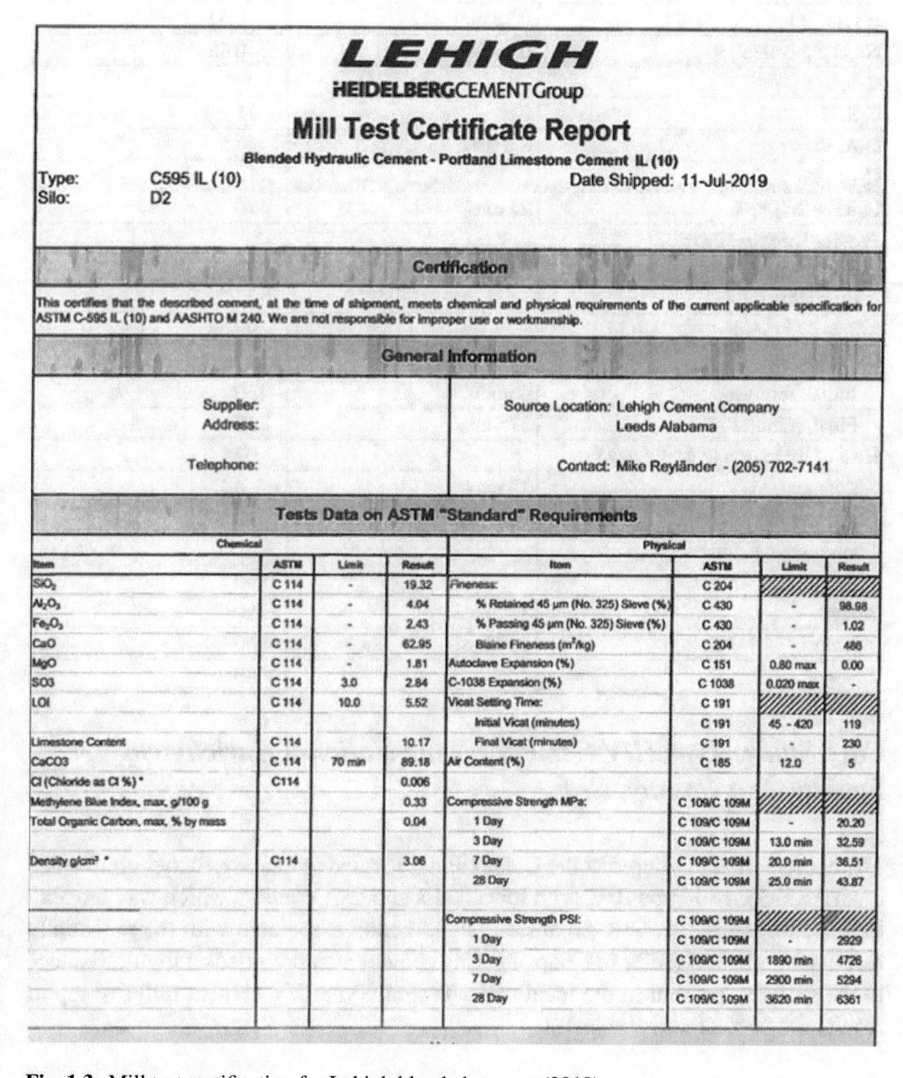

LEHIGH
HEIDELBERGCEMENTGroup

Mill Test Certificate Report

Blended Hydraulic Cement - Portland Limestone Cement IL (10)

Type: C595 IL (10)
Silo: D2
Date Shipped: 11-Jul-2019

Certification

This certifies that the described cement, at the time of shipment, meets chemical and physical requirements of the current applicable specification for ASTM C-595 IL (10) and AASHTO M 240. We are not responsible for improper use or workmanship.

General Information

Supplier:
Address:
Telephone:

Source Location: Lehigh Cement Company
Leeds Alabama
Contact: Mike Reyländer - (205) 702-7141

Tests Data on ASTM "Standard" Requirements

Chemical

Item	ASTM	Limit	Result
SiO_2	C 114	-	19.32
Al_2O_3	C 114	-	4.04
Fe_2O_3	C 114	-	2.43
CaO	C 114	-	62.95
MgO	C 114	-	1.81
SO3	C 114	3.0	2.84
LOI	C 114	10.0	5.52
Limestone Content	C 114		10.17
CaCO3	C 114	70 min	89.18
Cl (Chloride as Cl %) *	C114		0.006
Methylene Blue Index, max, g/100 g			0.33
Total Organic Carbon, max, % by mass			0.04
Density g/cm³ *	C114		3.06

Physical

Item	ASTM	Limit	Result
Fineness:	C 204		
% Retained 45 μm (No. 325) Sieve (%)	C 430	-	98.98
% Passing 45 μm (No. 325) Sieve (%)	C 430	-	1.02
Blaine Fineness (m²/kg)	C 204	-	486
Autoclave Expansion (%)	C 151	0.80 max	0.00
C-1038 Expansion (%)	C 1038	0.020 max	-
Vicat Setting Time:	C 191		
Initial Vicat (minutes)	C 191	45 - 420	119
Final Vicat (minutes)	C 191	-	230
Air Content (%)	C 185	12.0	5
Compressive Strength MPa:	C 109/C 109M		
1 Day	C 109/C 109M	-	20.20
3 Day	C 109/C 109M	13.0 min	32.59
7 Day	C 109/C 109M	20.0 min	36.51
28 Day	C 109/C 109M	25.0 min	43.87
Compressive Strength PSI:	C 109/C 109M		
1 Day	C 109/C 109M	-	2929
3 Day	C 109/C 109M	1890 min	4726
7 Day	C 109/C 109M	2900 min	5294
28 Day	C 109/C 109M	3620 min	6361

Fig. 1.3 Mill test certification for Lehigh blended cement (2019)

Table 1.3 Ash Grove Cement Type II/V mill test results

Requirements	ASTM C150 Type II	Ash Grove Type II/V
Chemical requirements		
SiO_2, %	20.0 min.	20.6
Al_2O_3, %	6.0 max	3.7
Fe_2O, %	6.0 max	3.2
CaO, %		63.7
MgO, %	6.0 max	3.2
SO_3, %	3.0 max	1.9
LOI, %	3.0 max	2.4
Insoluble residue, %	0.75 max	0.62
Na_2O equivalent, %	0.60 max	0.45
C_3S, %		59
C_2S, %		15
C_3A, %	8.0 max	4.5
C_4AF, %		10
$C_4AF + 2C_3A$, %	25 max	19
Physical requirements		
Blaine fineness, m^2/kg (air permeability)	280 min	402
Minus 325 mesh		99.60
Vicat set time		
Initial, minutes	45 min	130
Final, minutes	375 min	250
Heat of hydration (cal/g) 7 days		92
Air content, %	12 max	8.1
Autoclave expansion, %	0.80 max	0.04
Compressive strength, psi		
1 day		1970
3 days	1450 psi	3180
7 days	2470 psi	4960
28 days		6640

Case Study: Type II/V Cement for the National Enrichment Facility, Lea County, NM

The cement, manufactured at the Cemex Plant located in Odessa, Texas, conformed to an ASTM C150 Type II/V, with low alkalis and C_3A content, which was satisfactory in mitigating the ASR expansion phenomenon associated with the potentially reactive aggregates (refer to Chap. 2). This cement also provideds a high resistance to the sulfates, present in the local soils. Typical Type II/V cement mill test report results are presented in Table 1.2.

Table 1.4 Chickamauga LafargeHolcim Type II (MH) cement mill test results

Requirements	ASTM C150 Type II (MH)	CMTR Report Feb 2017
Chemical requirements		
SiO_2, %	20.0 min.	21.7
Al_2O_3, %	6.0 max	2.9
Fe_2O, %	6.0 max	4.5
CaO, %		64.1
MgO, %	6.0 max	3.1
SO_3, %	3.0 max	2.9
LOI, %	3.0 max	1.0
Insoluble residue, %	1.50 max	0.19
Na O equivalent, %	0.60 max	0.25
C_3S, %		62
C_2S, %		16
C_3A, %	8.0 max	0
C_4AF, %		14
$C_3S + 4.75{*}C_3A$, %		62
Physical requirements		
Blaine fineness, m^2/kg (air permeability)	260 min	303
Minus 325 mesh		82.5
Vicat set time		
Initial, minutes	45 min	89
Final, minutes	375 min	325
False set, %	50 min	72
Heat of hydration, cal/g	70	63
Air content, %	12 max	6.0
Autoclave expansion, %	0.80 max	−0.02
Mortar expansion, % (C11038)	0.02 max	0.000
Compressive strength, psi		
1 day		830
3 days	1450 psi	2000
7 days	2470 psi	3070
28 days		4870

Case Study: Type I/II Cement for a Montana Project

The cement, manufactured at the Ash Grove Montana City Plant, conformed to an ASTM C150 Type II, with low alkalis and C_3A content. ASTM C1260 expansion tests, made with Cemex cement and the selected aggregates, mitigated the ASR expansion phenomenon associated with the potentially reactive aggregates (refer to Chap. 2). The cement provides a high resistance to the sulfates, present in the local soils. Typical Type II/V cement mill test report results are presented in Table 1.3.

Case Study: Cement for a Mass Concrete Project at Chattanooga, TN

The cement, manufactured at the LafargeHolcim Plant, conforms to an ASTM C150 Type II (MH), with low alkalis and zero C_3A content. It mitigated the ASR expansion phenomenon associated with the potentially reactive aggregates (refer to Chap. 2). This cement provided a moderate resistance to the sulfates, present in the local soils.

Typical Type II (MH) cement mill test report results are presented in Table 1.4. This cement is being used for Chickamauga Lock at Chattanooga, TN. A typical Type I/II Buzzi Unicem cement mill test results is shown in Table 1.5.

Table 1.5 Chattanooga Buzzi Unicem Type I/II cement mill test results

Requirements	ASTM C150 Type I/II	CMTR Report Sep 2017
Chemical requirements		
SiO_2, %	20.0 min.	19.8
Al_2O_3, %	6.0 max	4.4
Fe_2O, %	6.0 max	3.2
CaO, %		62.5
MgO, %	6.0 max	3.2
SO_3, %	3.0 max	2.8
LOI, %	3.0 max	1.0
Insoluble residue, %	1.50 max	0.39
Na_2O equivalent, %	0.60 max	0.58
C_3S, %		61.6
C_2S, %		10.4
C_3A, %	8.0 max	6.2
C_4AF, %		9.8
$C_3S + 4.75*C_3A$, %		90.9
Physical requirements		
Blaine fineness, m^2/kg (air permeability)	260 min	389
Minus 325 mesh		95.5
Vicat set time		
Initial, minutes	45 min	89
Final, minutes	375 min	325
False set, %	50 min	61
Heat of hydration, cal./g		
Air content, %	12 max	7.2
Autoclave expansion, %	0.80 max	0.03
Compressive strength, psi		
1 day		2343
3 days	1450 psi	3912
7 days	2470 psi	4952
28 days		6507

Case Study: Type V Cement in South Carolina

Table 1.6 shows typical properties of Type V Lehigh cement.

Table 1.6 Type V Lehigh cement mill test results

Test requirements	ASTM C150-16 Type V	Results LEHIGH Type V, Leeds, AL 6/11/2015
Chemical requirements		
SiO_2, %		20.3
Al_2O_3, %		3.89
Fe_2O, %		4.1
CaO, %		63.47
MgO, %	6.0 max	3.45
SO_3, %	2.3 max	2.1
LOI, %	3.0 max	0.89
Insoluble residue, %	1.55 max	0.34
Na_2O equivalent, %	0.60 max	0.56
C_3S, %		66.3
C_2S, %		8.1
C_3A, %	5.0 max	3.4
C_4AF, %	25 max	12.5
Physical requirements		
Blaine fineness, m^2/kg (air permeability)	260 min	428
Minus 325 mesh, %		97.1
Gilmore set time		
Initial, minutes	60 min	174
Final, minutes	600 min	271
Air content, %	12 max	7
Autoclave expansion, %	0.80 max	
Fineness, Blaine	280	428
Autoclave expansion, %	0.8	0.09
Sulfate resistance, 14 days	0.04 max	
Compressive strength, psi		
1 day		2509
3 days	1160	3792
7 days	2180	4604
28 days	3050	5978

Case Study: Type I/II Argos Cement in South Carolina (Table 1.7)

Table 1.7 Type I/II Argos cement mill test results

Requirements	ASTM C150-16	ResultsARGOS Type II, SC 06/09/2016
Chemical requirements		
SiO_2, %		20.3
Al_2O_3, %	6.0 max	4.36
Fe_2O, %	6.0 max	3.04
CaO, %		63.60
MgO, %	6.0 max	2.32
SO_3, %	3.0 max	2.86
LOI, %	3.0 max	1.90
Insoluble residue, %	0.75 max	0.62
Na_2O equivalent, %	0.60 max	0.45
C_3S, %		61.1
C_2S, %		11.7
C_3A, %	8.0 max	6.3
C_4AF, %		9.2
$C_3S + 4.75*C_3A$, %	100 max	90
Physical requirements		
Blaine fineness, m^2/kg (air permeability)	260 min	410
Passing # 325 mesh (%)		96.3
Vicat set time		
Initial, minutes	45 min	108
Final, minutes	375 min	189
Heat of hydration (cal./g) 7 days		308
Air content, %	12 max	7
Autoclave expansion, %	0.80 max	−0.01
Fineness, Blaine	280	
Autoclave expansion, %	0.8	
Air content	12	7
Compressive strength, psi		
1 day		2359
3 days	1740 min	3885
7 days	2760 min	4602

1.3 Supplementary Cementitious Materials (SCM)

Supplementary cementitious materials (SCMs) include slag cement, Class F fly ash, and silica fume. SCMs have both hydraulic and pozzolanic properties.

ACI 318 recommends the quantity of fly ash, GGBFS or slag cement, and silica fume, subject to severe freezing and thawing exposure (Class F3) to a maximum of 50% of total cementitious materials by weight.

GGBFS Slag cement is available in three Grades (80, 100, and 120), based on slag activity index. It conforms to ASTM C989. Typically, slag cement is ground to a fineness exceeding that of Portland cement to attain increased activity at early ages. Slag cements are closest in chemical composition to Portland cement. Slag cements are rich in lime, silica, and alumina and have relatively more silica and less calcium than Portland cement. Slag Grade 100 (source: LafargeHolcim) is longer commercially available. Slag Grade 120 is commercially available from the Lehigh, Cape Canaveral source, and is being used for concrete mixtures. Slag Grade 120 is finer than slag Grade 100. Table 1.8 lists the Lehigh GGBF slag chemical and physical properties. A typical mill certification report for slag cement conforming to ASTM C989 is presented in Fig. 1.4.

Table 1.8 Slag cement chemical and physical properties

Requirements	ASTM C989	Average resultsLehigh Slag Grade 120 Cape Canaveral (June–July 2016)
Chemical requirements		
SiO_2, %		34.1
Al_2O_3, %		11.90
Fe_2O, %	6.0 max	0.64
CaO, %		40.28
MgO, %	6.0 max	6.3
SO_3, %	3.0 max	2.8
LOI, %	3.0 max	0.89
Na_2O equivalent, %	0.60 max	0.43
C_3S, %		65
C_2S, %		
C_3A, %		7
Physical requirements		
Blaine fineness, m^2/kg		539
Passing# 325 mesh (%)	80 max	99.36
Slag activity index 7 days (%)	90 min	106
Air content, %	12 max	3.0

Material Certification Report

LEHIGH
HEIDELBERGCEMENTGroup

Brand Name: Lehigh Slag Cement
Material: GGBFS
Type: ASTM C989 Grade 120

DATE: 01-Sep-2019
Silo # 611/612

General Information

Supplier:	Lehigh Cement Company	Source Location:	Lehigh Cement Company
Address:	575 Cargo Road Cape Canaveral, Florida 32920		575 Cargo Road Cape Canaveral, Florida 32920

The following information is based on monthly average test data. The data is typical of GGBFS shipped by Lehigh Cement Company. Cape Canaveral, FL Plant. Individual shipments may vary.

Test Data on ASTM "Standard" Requirements

Chemical (C989, Table 2)			Physical (C989, Table1)		
Item	Limit	Result	Item	Limit	Result
			+45 μm (No. 325) Sieve (%)	20 max	0.57
			Blaine Fineness (m2/kg)	-	522
Sulfide S (%)	2.5 max	0.9	Air Content (%)	12 max	1.8
			Expansion in Water (C-1038) (%)	0.020 max	0.015
Sulfate Ion - SO_3 (%)	NA	2.9	Slag Activity Index (SAI %)		
			Average of Last 5 Samples:		
			Avg 7 Day Index		99
Aluminum Oxide - Al2O3 (%)	NA	13.2	Avg 28 Day Index	115 min	125
			Current Samples:		
			7 Day Index		103
			28 Day Index	110 min	126

Test Data on CCRL Reference Cement

Chemical			Physical		
Item	Limit	Result	Item	Limit	Result
Total Alkalies as Na_2O (%)	0.60 - 0.90	0.78	Blaine Fineness (m2/kg)	-	382
C_3S	-	57	Compressive Strength MPa (psi)		
C_2S	-	15	7 Day	-	4412
C_3A	-	7	28 Day	34.5 (5000) min	39.0 (5650)
C_4AF	-	8			

Optional Test Data

Chemical			Physical		
Item	Limit	Result	Item	Limit	Result
% Total Alkalies	-	0.40	Specific Gravity (Latest Result)	-	2.86
%Cl (Chloride)	-	<0.01			

Certification Statement

Lehigh Slag Cement meets Section 929-1 and 929-5 of FDOT Specifications

Lehigh Cement Office Cape Canaveral, FL - (321) 323-5032

Fig. 1.4 Typical mill certification report for Lehigh slag cement (Sep 2019)

ACI 233, "Slag Cement in Concrete and Mortar," discusses the effectiveness of the slag cement in concrete. Slag cement improves sulfate resistance of concrete, based on the tricalcium aluminate and the alumina content of the slag cement. The permeability of concrete containing slag cement is greatly reduced compared to concrete containing Portland cement only, with decreased permeability as the proportion of slag is increased. The excess silica in slag reacts with the calcium hydroxide and alkalis released during the cement hydration, leading to C-S-H filling concrete pores. A reduction in pore size leads to reduction in permeability and enhances resistance to penetration of chlorides. Due to the complexity of the reacting system, ACI 233 recommends that direct performance evaluation of the slag cements for workability, strength characteristics, and durability should be made using materials intended for the work.

Fly Ash Fly ash is finely divided residue created from the combustion of bituminous and anthracite coals, with a low calcium oxide percent. Fly ash is a pozzolan, and the siliceous and aluminous materials in the fly ash alone possess little cementitious value. Fly ash particles are mostly spherical with a mean particle size diameter similar to that of Portland cement. Fly ash is intended to mitigate sulfate attack, reduce permeability, and improve durability of concrete. ACI Committee 232 Report provides the uses of fly ash in concrete. ASTM C618 covers the requirements of coal fly ash and raw or calcined natural pozzolans for use in concrete.

Table 1.9 lists the chemical and physical properties of Class F fly ash. It should be noted that incorporating fly ash Class F, which has pozzolanic properties only, in concrete would slow the rate of strength development in concrete. Class C fly ash, in addition to pozzolanic properties, also has some cementitious properties that may not affect the rate of strength development.

Fly ash and natural pozzolans must be tested, prior to use, including all optional requirements for effectiveness in controlling the alkali-silica reaction, drying shrinkage, and uniformity. All fly ash used in concrete should be from a single source. Fly ash and its use should be in accordance with the requirements of ACI standards, unless otherwise specified. ACI 318 limits fly ash and pozzolans to a maximum of 25% by weight of cement for concrete exposed to very severe freezing and thawing (Class F3). Refer to Table 2.1.

Before fly ash is accepted for use, it should be tested in combination with the specific cement and aggregates being used. All fly ash used for the project must come from one source.

Silica Fume Silica fume must conform to ASTM C1240. Silica fume should be furnished as a dry, densified material. Silica fume should be limited to a maximum of 10% by weight of cement. All silica fume used for the project must come from one source.

Table 1.9 Class F fly ash chemical and physical properties

Test requirements	ASTM C150-16 Type V	Results SEFA August 4 2016 Results
Chemical requirements		
SiO_2, %		53.5
Al_2O_3, %		27.2
Fe_2O, %		10.0
CaO, %		2.1
MgO, %	6.0 max	1.0
SO_3, %	2.3 max	0.13
LOI, %	3.0 max	0.7
Insoluble residue, %	1.55 max	0.34
Na_2O equivalent, %	0.60 max	1.91
Physical requirements		
Retained on Minus 325 mesh, %	34	24.0

Water The water for washing aggregate and mixing and curing concrete should be potable. It should be clean and free from significant quantities of organic matter, alkali, salts, oil, sediments, and other impurities that might reduce the strength and durability or otherwise adversely affect the quality of the concrete. Potable drinking water is satisfactory for concrete. Water should not contain more than 650 parts per million of chlorides as Cl nor more than 1000 parts per million of sulfates as SO4.

ACI 318 allows non-potable water for use in mixing concrete, subject to qualification in accordance with ASTM C1602. When tested in accordance with ASTM C109, the water should cause no more than a 25% change in setting time or a 5% reduction in 14-day compressive strength of the mortar cube when compared to results using distilled water.

Water for curing should not contain any impurities in a sufficient amount to cause discoloration of the concrete or mortar or produce etching of the surface.

Seawater: Seawater is not suitable for use in making steel-reinforced concrete and prestressed concrete due to high risk of steel corrosion.

Wash Water: Wash waters may be reused as mixing water in concrete if they satisfy the limits in Table 1.10.

1.4 Concrete Aggregates

General Aggregates are the granular fillers in cement that can occupy as much as 60–75% of total volume. They influence the properties of freshly mixed and hardened concrete and economy. Concrete aggregates are mixtures of naturally occurring minerals and rocks. The most desirable aggregates used in concrete are hard, dense, well graded with durable properties to resist weathering.

Aggregate Characteristics There are many different types of rocks used as aggregate. Hardness often varies even within the same classification of rock. For example, granite varies in hardness and friability. Figure 1.5 shows an aggregate map of the USA and Canada, with regional variations from soft to hard rock formations.

Table 1.10 Chemical limits for wash water used as mixing water (ASTM C94)

Chemical or type of construction	Maximum concentration parts per million (ppm)	Test method
Chloride, as Cl		ASTM D 512
Prestressed concrete or concrete in bridge decks	500	
Other reinforced concrete	1000	
Sulfate, SO4	3000	ASTM D 516
Alkalis, as ($Na_2O + 0.658\ K_2O$)	600	
Total solids	50,000	AASHTO T 26

Fig. 1.5 Aggregate map of the USA and Canada

Hardness of the Aggregate Mohr's scale is frequently used to measure hardness. Values of hardness are assigned from 1 to 10. A substance with a higher Mohr's number scratches a substance with a lower number – higher Mohr's scale numbers indicate harder materials. The scale below shows how some common minerals fall into Hoh's scale range.

Mohr's range	Description	Aggregate
8–9 Pebble	Very hard	Flint, chert, trap rock, basalt, river rock
6–7 Quartz, trap rock	Hard	Hard river rock, hard granite, basalt
4–5	Medium hard	Medium hard granite, some river rock
3–4 Marble	Medium	Dense limestone, sandstone, dolomite
2–3	Medium soft	Soft limestone

Aggregates containing soft and porous rocks, including certain types of cherts, should be avoided due to low resistance to weathering and potential surface defects. Since aggregates are the major constituents of concrete, they greatly influence the quality of concrete. Physical characteristics of aggregates such as gradation, absorption, and specific gravity have significant effects on quality. Aggregate size and shape can have significant effects on water demand of concrete and durability of concrete.

ASTM C33 defines the minimum requirements for grading and quality of fine and coarse aggregates for use in concrete. It includes the permissible amounts of deleterious substances, soundness, abrasion resistance, and alkali reactivity, demonstrated through tests and/or satisfactory service records. Aggregates consisting of materials that can react with alkalis in cement and cause excessive expansion, cracking, and deterioration of concrete mix should preferably never be used. Therefore, it is required to test aggregates to know whether there is presence of any such constituents in aggregate or not. If there is no economic alternative, such materials may be used if other measures such as low-alkali cement, fly ash, or GGBF slag are used to minimize the impact.

The Size and Shape of the Aggregate ASTM C33 grading requirements permit a wide range of grading sizes, from #1 size (3–1/2 to 1–1/2 inch) to #8 size (3/8 in. to No. 8). It allows a selection of suitable coarse aggregate size based on job requirements.

The maximum size of coarse aggregate used in concrete is based on economy and the type of structure. The larger grading sizes are used for mass concrete in dams and structural monoliths, which require placements by buckets and conveyer belts, and require nominal reinforcement, to control cracking, while the smaller grading sizes are used for conventional reinforced concrete structures, including buildings and pavements, that require placement by pumping method.

It should be noted that the size and shape of aggregate particles influence the properties of freshly mixed concrete more as compared to those of hardened concrete.

For the preparation of economical concrete mixtures, one should use the largest coarse aggregates feasible for the structure.

The development of hard bond strength between aggregate particles and cement paste depends upon the surface texture, surface roughness, and surface porosity of the aggregate particles. If the surface is rough but porous, maximum bond strength develops. In porous surface aggregates, the bond strength increases due to setting of cement paste in the pores.

Specific Gravity (Relative Density) and Voids Specific gravity (relative density) is primarily of two types, apparent and saturated surface dry basis (SSD), calculated by the following equations:

$$\text{Apparent Specific Gravity} = A/\left(A \quad C\right),$$

$$\text{SSD Specific Gravity} = B/\left(B \quad C\right)$$

where:

A = mass of oven-dried sample in air
B = mass of SSD sample in air
C = apparent mass of saturated sample in water

Specific gravity is a means to deciding the suitability of the aggregate. Low specific gravity generally indicates porous, weak, and absorptive materials, whereas high specific gravity indicates materials of good quality. Specific gravity of major aggregates falls within the range of 2.6–2.9. Specific gravity values are used while designing concrete mixture.

Note: The standard test methods for the determination of specific gravity of coarse aggregate (AASHTO T85; ASTM C127) are essentially the same, except for the required time in which a sample of aggregate is submersed in water to essentially fill the pores – the AASHTO method (AASHTO T85) requires the sample be immersed for a period of 15–19 h, while the ASTM method (ASTM C127) specifies an immersed period of 24 ± 4 h.

Voids: The empty spaces between the aggregate particles are known as voids. The volume of void equals the difference between the gross volume of the aggregate mass and the volume occupied by the particles alone. The minute holes formed in volcanic rocks during solidification of the molten magma, due to air bubbles, are known as pores. Rocks containing pores are called porous rocks.

Water absorption may be defined as the difference between the mass of oven dry aggregates and the mass of the saturated aggregates with surface dry conditions. Depending upon the amount of moisture content in and on the aggregates, it can exist in any of the four conditions.

- Oven dry aggregate (having no moisture)
- Air Dry aggregate (contain some moisture in its pores)
- Saturated surface dry aggregate (pores completely filled with moisture but no moisture on surface)
- Moist or wet aggregates (pores are filled with moisture and also having moisture on surface)

Fineness Modulus Fineness modulus (FM) is an empirical factor obtained by adding the cumulative percentages of aggregate retained on each of the standard sieves ranging from 80 mm to 150 micron and dividing this sum by 100. Fineness modulus is generally used as a gauge of how coarse or fine the aggregate is. Higher fineness modulus value indicates that the aggregate is coarser, and small value of fineness modulus indicates that the aggregate is finer.

Specific Surface: The surface area per unit weight of the material is termed as specific surface. This is an indirect measure of the aggregate grading. Specific surface increases with the reduction in the size of aggregate particle. The specific surface area of the fine aggregate is much more than that of coarse aggregate.

Deleterious Materials: Aggregates may contain any harmful material in such a quantity so as to affect the strength and durability of the concrete. Such harmful materials are called deleterious materials. Deleterious materials may prevent proper bond and affect setting time, strength, and durability.

Abrasion Resistance: The abrasion value gives a relative measure of resistance of an aggregate to wear when it is rotated in a cylinder along with some abrasive charge and is a measure of the durability of the aggregate.

All concrete aggregates must be in accordance with the requirements of ASTM C33. Typical aggregate characteristics and test requirements and acceptance criteria are presented in Table 1.11.

Aggregate Source Fine and coarse aggregate should be obtained only from a source shown to be in compliance with the performance requirements. The contractor should notify the owner's representative in writing not less than 14 days prior to use, naming the source of fine and coarse aggregate. The source of fine or coarse aggregate should not be changed during the course of the work unless the new source is properly investigated and its aggregate properties determined and documented.

Table 1.11 Aggregate characteristics and test requirements

Characteristics	Significance	ASTM test reference	Acceptance criteria
Abrasion resistance	Durability of surface	C131 and C535	Limit loss to 30%
Freezing and thawing resistance	Surface defects and scaling	C666 (Procedure A)	Limit weight loss to 50% after 500 revolutions
Particle shape and texture	Workability	C295	Limit flat and elongated particles to 3%
Grading	Workability and economy	C136	ASTM C33
Specific gravity	Concrete mixture proportions	C127 and C128	2.4 minimum
Absorption	Concrete mixture proportions	C127 and C128	2.5% max.
<200 sieve material	Deleterious materials	C117, C123. C142,	3% max.
Resistance to alkali reactivity	Volume stability and soundness	C1260, C1293, C1567	Limit length change to 0.10%

Fine and coarse aggregates should be tested for detecting deleterious alkali-aggregate reactivity. Refer to Chap. 3.

Fine Aggregate (Sand) Fine aggregate for concrete should consist of natural or manufactured sand, or a blend of the two fine aggregates, and should in all cases be washed. The contractor should do all sorting, crushing, screening, blending, washing, and other operations necessary to ensure that the material complies with the requirements. If the finer particles from the crushed coarse aggregate are to be mixed with the sand from natural deposits, the two components should be uniformly blended before washing or screening to ensure a combined product of constant composition. Table 1.12 shows ASTM C33 grading requirements for fine aggregate.

Fine aggregate should be of such quality as to develop a relative mortar strength of not less than 9% when tested in accordance with the requirements of ASTM C87.

Washed or saturated sand should be allowed to drain for at least 24 hours to a uniform moisture content before batching. Dry sand should be moistened before handling to prevent segregation.

1.5 Coarse Aggregate

General Coarse aggregate should be gravel, crushed rock, or a combination thereof and should be composed of strong, hard, clean, durable uncoated rock, free from alkali. Coarse aggregate should be free of organic and other deleterious matter.

Coarse aggregate should be washed and, if necessary, uniformly moistened just before batching. Coarse aggregate should be stored in separate batching bins and batched as required in accordance with the combined grading requirements.

For reinforced concrete members, ACI 318 Code requires that nominal maximum size of aggregate shall be not larger than:

Table 1.12 ASTM C33 fine aggregate grading requirements (percent passing)

Property	ASTM C33	AASHTO M6
Gradation (sieve size)		
3/8 inch (9.5 mm)	100	100
#4 (4.75 mm)	95–100	95–100
#8 (2.36 mm)	80–100	80–100
#16 (1.18 mm)	60–80	60–80
#30 (600 μm)	25–50	35–60
#50 (300 μm)	5–30	10–30
#100 (150 μm)	0–10	2–10
#200 (75 μm)		

ASTM C33 requires that FM of sand should be between 2.3 and 3.1. The FM should not vary from than 0.2 from the typical value of the selected aggregate source to assure uniformity of concrete during production

- One-fifth the narrowest dimension between sides of forms
- One-third the depth of slab
- Three-fourth the minimum clear spacing between reinforced bars or wires, bundles of bars, or tendons

The nominal maximum size of coarse aggregate, by member thickness, is recommended below:

Thickness	Nominal maximum size
Inches	*Aggregate (ASTM C33) inches*
Less than 8	3/8 inch (No. 8)
8–12	1 (No. 57)
>12	1–1/2 (No. 467)

Case Study: NEF Project, Lea County, New Mexico

Fine Aggregate (Sand): Gradation results from various pits are shown in Table 1.13.

Coarse Aggregate: Valentine Pit coarse aggregate source was selected since it had the highest specific gravity, and the lowest absorption values, shown in Table 1.14.

Table 1.13 NEF Project fine aggregate grading requirements and test results (percent passing)

Sieve size	ASTM C33 percent passing	Imperial	Slaton	Grand Falls	Valentine	Grisham
3/8 inch (9.5 mm)	100	100	100	100	100	100
#4 (4.75 mm)	95–100	98	100	91	98	99
#8 (2.36 mm)	80–100	87	88	76	80	90
#16 (1.18 mm)	50–85	76	71	64	66	80
#30 (600 μm)	25–60	61	58	51	52	64
#50 (300 μm)	5–30	25	25	22	31	29
#100 (150 μm)	0–10	4	4	4	9	4
#200 (75 μm)	0–3	1.1	0.6	1	3	2.5
FM	2.3–3.1	2.49	2.54	2.92	2.64	2.34
Specific gravity		2.49	2.61–2.62	2.62–2.64	2.643	2.607
Absorption. %			1.5	0.68–1.1		
Organic impurities (ASTM C40)			None	None		
Soundness (ASTM C88), %			9	6		

Table 1.14 Coarse aggregate test summary (percent passing)

Sieve size	Specification size No. 57 (1 in. to No. 4)	SE-2 washed	Valentine Pit (Lafarge)	Wallach	Kiewit #57
1–1/2 in	100	100	100	100	100
1 in (25 mm)	95–100	100	100	100	100
1/2 (12.5 mm)	25–60	26	21	50	43
3/8 (9.5 mm)					
#4 (4.75 mm)	0–10	1	1	3	1
#8 (2.36 mm)	0–5				
#200 (75 μm)	1% max		0.6		
Specific gravity, SSD		2.744	2.758	2.411	
Absorption, %		0.94	0.9	4.5	
Soundness, % loss			0.68		
Los Angeles abrasion, % loss			19.1		
Flat and elongated particles, %			11.8		
Clay lumps, %			None		
Lightweight pieces, %			**0.2**		

Case Study: Coarse Aggregates for Chickamauga Lock, TN (Table 1.15)

Table 1.15 ASTM C33 grading requirements for Chickamauga coarse aggregates (percent passing)

Property	Size #8 (3/8 in. to #8)	Size #67[a] (3/4 in. to #4)	Size #467[b] (1–1/2 to No. 4)	Size #357[c] (2 inch to #4)
Gradation (sieve size)				
2–1/2 inch (63 mm)				100
2 inch (50 mm)			100	95–100
1–1/2 in (37.5 mm)			95–100	
1 in (25 mm)		100	20–45	35–70
3/4 in (10 mm)		90–100	35–70	
1/2 (12.5 mm)	100			10–30
3/8 (9.5 mm)	85–100	20–55	10–30	
#4 (4.75 mm)	10–30	0–10	0–5	0–5
#8 (2.36 mm)	0–10	0–5		
#16 (1.18 μm)	0–5			
#200 (75 μm)	1% max	1% max	1% max	1% max

[a]For concrete thickness less than 12 inches

[b]For structural concrete and thickness ranging from 12 to 24 inches, the maximum nominal size of aggregate shall be 1–1/2 inches

[c]For mass concrete, the maximum nominal aggregate size shall be 3 inches

Case Study: Aggregates for a South Carolina Project

Aggregate Sources: All concrete aggregates conformed to the requirements of ASTM C33.

Fine Aggregate: Natural river material Beech Island, SC.

Coarse Aggregate: Dogwood, crushed igneous rock with low absorption (Tables 1.15 and 1.16).

Fine aggregate (sand) and coarse aggregate combined gradation (#67, #57, #467) are shown in Tables 1.17, 1.18, and 1.19.

Aggregate Optimization The actual combined aggregate gradations for sizes #67 (3/4″), #57 (1″), and #467 (1.5″), using 0.45 power graph, are shown in the 0.45 power graph in Fig. 1.6. The combined aggregate gradations for each aggregate size require optimization to improve workability and to minimize segregation.

Historically, the 0.45 power chart has been used to develop uniform gradations for asphalt mixture designs; however, it has now been widely used to develop uniform gradations for Portland cement concrete mixture designs. Some reports have circulated in the industry that plotting the sieve opening raised to the 0.45 power may not be universally applicable for all aggregates. In this paper the validity of 0.45 power chart has been evaluated using quartzite aggregates. Aggregates of different sizes and gradations were blended to fit exactly the gradations of curves raised to 0.35, 0.40, 0.45, 0.50, and 0.55. Five mixtures, which incorporated the aggregate gradations of the five power curves, were made and tested for compressive

Table 1.16 Quality requirements for Chickamauga coarse aggregates

Description	Desired value
Specific gravity, SSD (ASTM C128)	2.65
Absorption, % (ASTM C128)	2.5
Magnesium sulfate soundness, % loss (ASTM C88)	<7%
Durability factor ASTM C666 and CRD-C114(Procedure A – 12 cycles per day)	>50%
Los Angeles abrasion, % loss ASTM C131 and C535	<30%
Flat and elongated particles, % ASTM D4791)	<12
Clay lumps, % ASTM C142	<1
Material finer than No.200 sieve, % (ASTM C117)	<1
Lightweight pieces (chert with less than 2.4 S.G.) (ASTM C123)	<1.0%
Petrographic examination, ASTM C295 Argillaceous Chert Dolomite Insoluble residue Strained quartz Ferrous materials	<3% <10% Chert <5% Chert with Chalcedony <10% <10% <20% <2%
Alkali-silica reactivity, ASTM C1260 ASTM C1567 (with job mix design)	<0.10% at 14 days

Table 1.17 #67 coarse aggregate and sand combined gradation

	#67 coarse aggregate	Sand	Combined gradation	
Sieve size	% passing	% passing	% passing	% retained
1.5 inch	100	100	99.99	0.01
1 inch	100	100	99.99	0.00
0.75 inch	96.53	100	97.85	2.14
0.5 inch	69.14	100	80.93	16.92
0.375 inch (9 mm)	39.36	100	62.53	18.40
No. 4 (4.75 mm)	5.46	99.47	41.39	21.14
No. 8 (2.36 mm)	2.24	94.76	37.60	3.79
No. 16 (1.18 mm)	2.24	75.75	30.34	7.26
No. 30 (600 μm)	2.24	43.28	17.93	12.41
No. 50 (300 μm)	0	14.17	5.42	12.51
No. 100 (150 μm)	0	1.8	0.69	4.73
No. 200 (75 μm)	0	0	0.00	0.69
Pan	0	0	0.00	0.00
Fraction	0.62	0.38	1.00	

Note: The absorption of #67 stone is 0.96%. The absorption of fine aggregate (sand) is 0.7%

Table 1.18 #57 coarse aggregate and sand combined gradation

Sieve size	#57 coarse aggregate% passing	Fine aggregate% passing	Combined gradation% passing	Combined gradation % retained
1.5″	100	100	100	0
1″	99.9	100	99.93	0.07
0.75″	85.5	100	90.56	9.37
0.5″	37.9	100	59.56	31.00
0.375 (9 mm)	14.6	100	44.39	7.17
No. 4 (4.75 mm)	3.3	99.47	36.85	7.54
No. 8 (2.36 mm)	1.8	94.76	34.23	2.62
No. 16 (1.18 mm)	0	75.75	26.42	7.81
No. 30 (600 μm)	0	43.28	15.10	11.32
No. 50 (300 μm)	0	14.17	4.94	10.16
No. 100 (150 μm)	0	1.8	0.63	4.31
No. 200 (75 μm)	0	0	0	0.63
Pan	0	0	0	
Total	1850	991	2841	
Fraction	0.651	0.349		

Note: The absorption of #57 stone is 0.50%. The absorption of fine aggregate (sand) is 0.7%

Table 1.19 #467 coarse aggregate and sand combined gradation

Sieve size	#4 coarse aggregate % passing	#67 coarse aggregate % passing	Fine aggregate % passing	Combined gradation % passing	Combined gradation % retained
2″	100	100	100	100	0
1.5″	94.2	100	100	99.07	0.93
1″	48.4	100	100	91.74	7.33
0.75″	10.4	96.53	100	84.01	7.73
0.5″	3.5	69.14	100	69.85	4.16
0.375″ (9 mm)	1.8	39.36	100	55.10	14.75
No. 4 (4.75 mm)	0.8	5.46	99.47	38.89	14.21
No. 8 (2.36 mm)	0	2.24	94.76	35.51	3.38
No. 16 (1.18 mm)	0	2.24	75.75	28.60	6.91
No. 30 (600 μm)	0	2.24	43.28	16.80	11.80
No. 50 (300 μm)	0	0	14.17	5.15	11.65
No. 100 (150 μm)	0	0	1.8	0.65	4.50
No. 200 (75 μm)	0	0	0.8	0.29	0.36
Pan	0	0	0	0	0.29
Subtotal (lbs.)	455	1355	1034	2844	

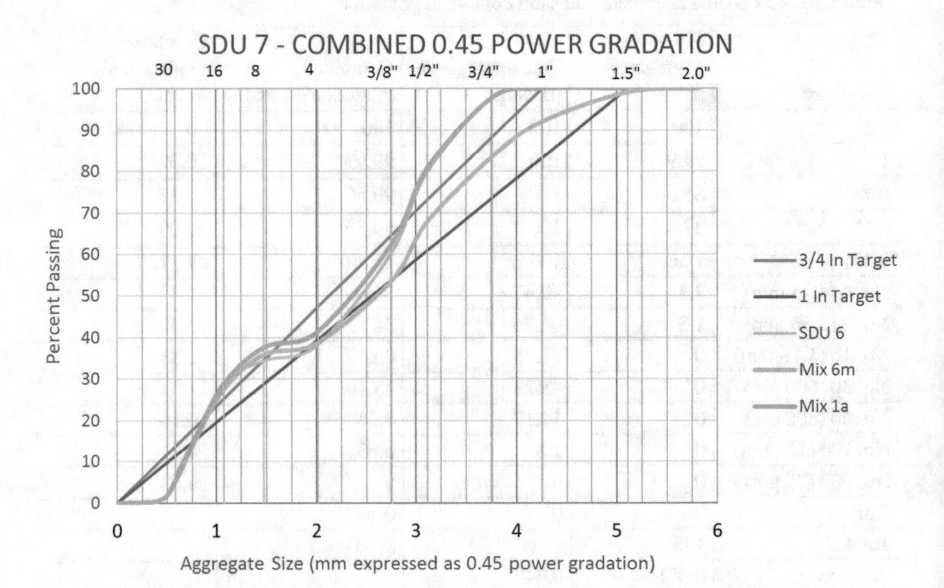

Fig. 1.6 Combined 0.45 power gradation for #57 and #467 Aggregates for a SC project

strength and flexural strength. A control mixture was also made whose aggregate gradations did not match the straight-line gradations of the 0.45 power curve. This was achieved by using a single size aggregate and sand. The water-cement ratio and the cement content were kept constant for all the six mixtures. The results showed that the mixture incorporating the 0.45 power chart gradations gave the highest strength when compared to other power charts and the control concrete. Thus, the 0.45 power curve can be adopted with confidence to obtain the densest packing of aggregates, and it may be universally applicable for all aggregates [1].

Figures 1.7 and 1.8 provide the workability and coarseness factors for #57 and #467 combined gradation. The results indicate that the #467 coarse aggregate gradation, a blend of two aggregate fractions (#4, #67), offered coarseness factor of 71.8% and a workability factor of 35.7% that was better than the #57 combined gradation, as shown in Figs. 1.5 and 1.6.

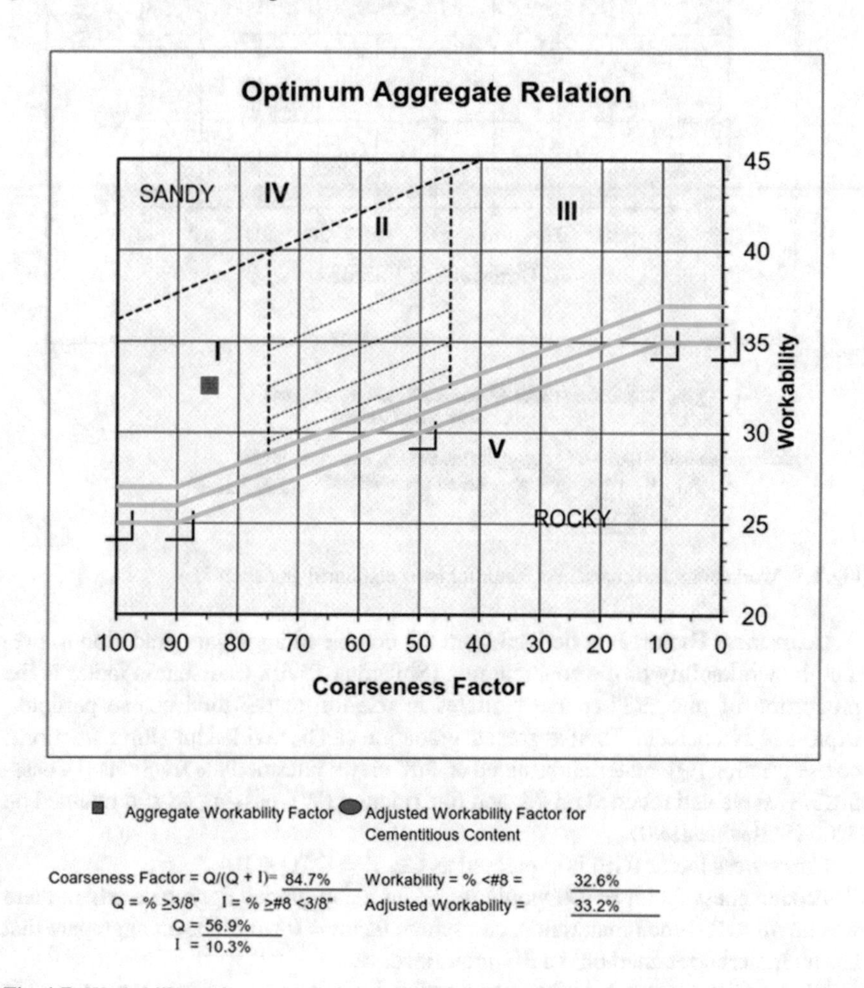

Fig. 1.7 Workability and coarseness factor for #57 combined gradation

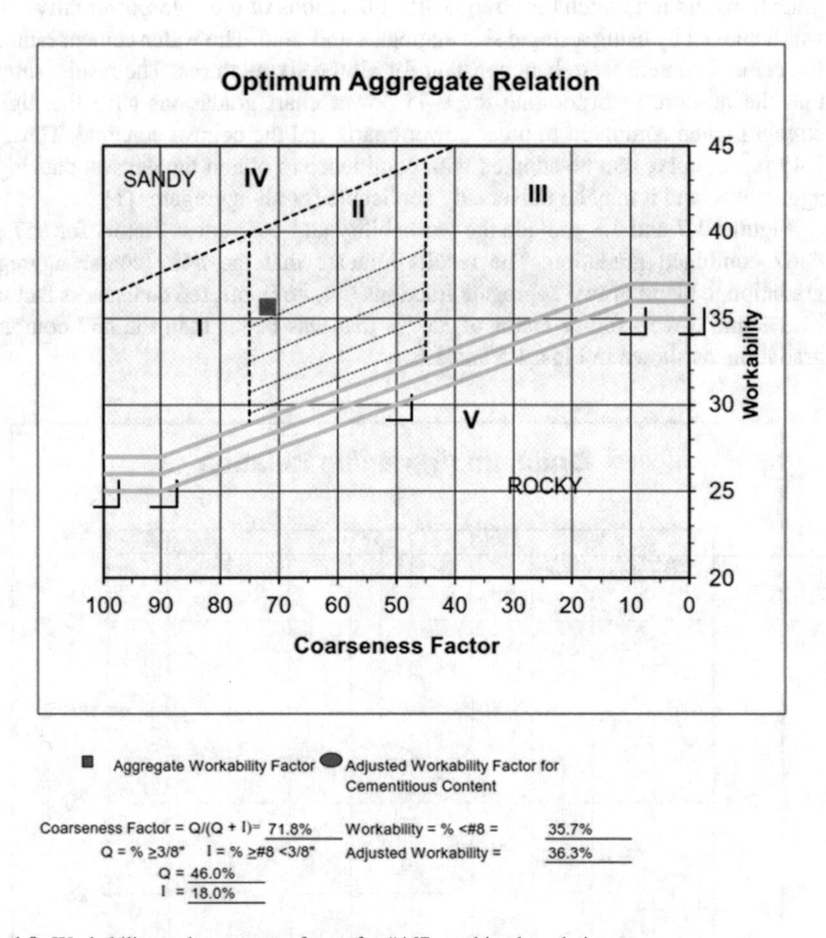

Fig. 1.8 Workability and coarseness factor for #467 combined gradation

Coarseness Factor: It is derived from the combined aggregate gradation to predict the workability of the concrete mix (Shilstone 1990). Coarseness factor is the proportion of plus 3/8″ coarse particles in relation to the total coarse particles, expressed as a percent. Total aggregate gradation can be divided into three fractions: coarse fraction (Q), materials retained on 3/8″ sieve; intermediate fractions (I), passing 3/8″ sieve and retained on #8; and fine fraction (W), passing #8 and retained on #200 (Shilstone 1990).

Coarseness factor (CF) is expressed as CF, % = Q/(Q + I).

A coarseness factor = 100 would represent a gap-graded aggregate where there was no #8 to 3/8-inch material. A coarseness factor = 0 would be an aggregate that has no material retained on the 3/8-inch sieve.

Workability Factor: It is the percent of the combined aggregate gradation that passes the No. 8 sieve. The Coarseness Factor Chart is based upon 6.0 sacks (564 pounds) of cementitious materials per cubic yard (335 kg/m^3); it needs to be adjusted in order to account for different cementitious amounts in a concrete mixture. When the amount of cement exceeds 6.0 sacks, the workability factor is adjusted plus 2.5% per sack of cement equivalent. When the amount of cement is below 6.0 sacks, the workability factor is adjusted minus 2.5% per sack of cement equivalent.

1.6 Admixtures for Concrete

Air-Entraining Admixtures

General Air entrainment refers to the introduction of large quantities of tiny air bubbles in the concrete matrix. The main reason for air entrainment is to improve the durability of the concrete to freeze-thaw degradation in the presence of moisture by intersecting the capillary network at many locations. Damage to concrete by freezing and thawing is caused when ice forms in the capillaries to create a pressure greater than tensile strength of the cement paste. Under a well-dispersed air-void system, the travel distance to move water or ice is minimized, resulting in reduced pressure differential.

ASTM C125 defines the terms "entrapped air" and "entrained air": the size of air voids below 0.04 inch (1 mm) is considered "entrained air" and entrapped above this nominal limit.

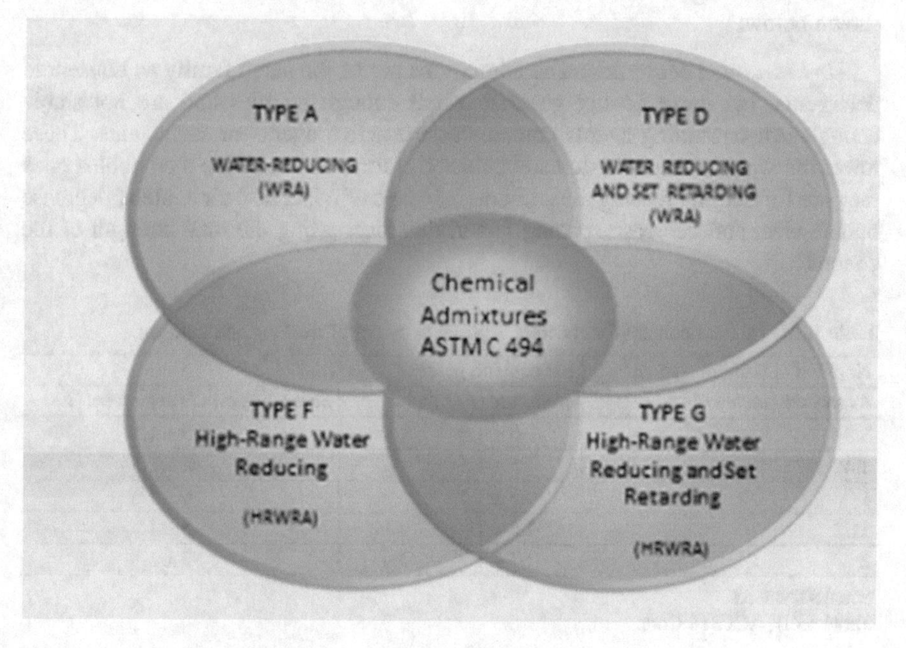

Fig. 1.9 Types of ASTM C494 chemical admixtures for concrete

The Air-Void System: As un-reacted water in concrete matrix freezes, it expands 9% by volume on phase change. This internal volume expansion causes internal stresses in the matrix. It can generate cracks in the concrete, which may allow water to infiltrate and the process can get progressively worse. It can lead to significant degradation of the concrete.

The formation of ice in the pore spaces generates pressure on any remaining unfrozen water. Introducing a large quantity of air bubbles provides a place for this water to move in to relieving the internal pressure. What is desired is to generate very many small air bubbles well distributed throughout the matrix rather than a smaller number of larger bubbles.

It's been determined that the optimum air content for frost protection is about 9% by volume of the mortar fraction. With respect to the concrete volume, the air content should be in the range of 4–8% by volume. The concrete normally has entrained air; the admixture increases the total volume of the air voids by 3–4% of the concrete volume.

Total air content is only a part of the formula for frost resistance. The nature of the entrained air is equally important. The critical parameter of the air-entrained paste is the spacing factor (max distance from any point in the paste to the edge of a void). It should not exceed 0.2 mm; the smaller the spacing factor, the more durable the concrete.

Air-Entraining Admixture Air-entraining admixture should conform to ASTM C260. ACI 318 Code requires concrete exposed to freezing and thawing exposure to be air-entrained, as shown in Table 1.20. The air-entraining shall be added to provide the percentage of air entrainment in the concrete as discharged from the mixer shown below:

Air-Entraining Materials: The admixtures are of the same family as household detergents, but these do not generate small enough bubbles and are not stable enough. Air-entraining agents contain surface-active agents or surfactants. These lower the water surface tension so bubbles can form and stabilize the bubbles once they are formed. Increasing the admixture dosage will increase air content, decrease bubble size, and decrease spacing factor, thus decreasing the total strength of the concrete.

Table 1.20 Total air content for concrete exposed to cycles of freezing and thawing

Nominal maximum size, in.[a]	Air content, percent[b]	
Aggregate (inches)	*Moderate (F1)*	*Severe (F3) and very severe (F3)*
3/8	6 ± 1.5	7.5 ± 1.5
3/4	5 ± 1.5	6 ± 1.5
1	4.5 ± 1.5	6 ± 1.5
11/2	4.5 ± 1	5 ± 1
2	4 ± 1	5 ± 1

[a]See ASTM C33
[b]Table 4.4.1, ACI 318 Code

Air content of concrete is determined in accordance with ASTM C173 and C231.

Effect of Air on Concrete Properties The air entrainment is also useful for improving the workability of concrete during placement. A large number of microscopic small voids, well dispersed throughout the cement paste, will generally increase workability at any water content. In proportioning concrete mixtures, one can take advantage of this increased workability by reducing the water content by 6–12%, depending on the aggregate gradation and other factors, while maintaining slump.

The use of fly ash in concrete does have a potential impact on the air entrainment of concrete due to its carbon content. Fluctuations in chemistry of the fly ash, such as loss of ignition (LOI), between shipments, will result in fluctuations in concrete air content and require monitoring.

The use of water-reducing admixtures (WRA) and high-range water-reducing (HRWR) admixtures does also influence the air entrainment of concrete. This dosage rate of the admixtures to offset the air entrainment with time needs to be verified by testing.

Typical effects of entrained air on concrete properties include:

- Increase workability and cohesiveness of fresh concrete.
- Considerable reduction in bleeding and segregation.
- Decreased strength (10–20% for most air-entrained concrete).
- Increased durability (up to ~7% air).
- Characteristics result in more impermeable concrete and a better overall resistance to aggressive agents (i.e., sulfates).

Air-entraining admixtures are manufactured by several manufactures and conform to ASTM C260.

The dosage rate of the air-entraining admixture must be established, prior to production, by trial batches of concrete simulating the anticipated range of the site conditions. Typical factors which influence the dosage of an air-entraining admixture include temperature, cementitious materials' type and amount, water/cementitious material ratio, fine aggregate gradation, sand-aggregate ratio, admixtures, site conditions, slump, mixing and delivery time, and placement methods.

The established dosage rate of the air-entraining admixtures may require adjustments during actual production of concrete, dictated by field conditions and placement methods. A high air will produce a reduction of strength that cannot be recovered with age, whereas a low air content will reduce the durability characteristics of concrete in a cold climate.

Non-air-entrained concrete, on the other hand, if it remains dry or contains a small amount of moisture, sustains no damaging effects from freezing and thawing. Non-air-entrained concrete with a good-quality aggregate, a high cementitious content (564 lb/cy), a compressive strength of 4500 psi, and low water-to-cementitious material ratio (0.40) develops good resistance to freezing and thawing primarily because of its relative high density and water tightness which reduce the free water available to the capillary system. Durability and freeze-thaw resistance capability of a non-air-entrained concrete mix may be evaluated in accordance with ASTM C666.

1.7 Chemical Admixtures

General There are water-soluble compounds added primarily to control setting and early hardening of fresh concrete or to reduce the water requirements. Water-reducing or water-reducing and water-retarding admixtures for concrete should be in accordance with the ASTM C494, Types A to G. In addition, all water-reducing admixtures should be in accordance with the following:

Admixture should be a polymer type, a specially manufactured derivative of a hydroxylated carboxylic acid, or a lignosulfonate base.
Admixture should contain no calcium chloride.
Admixture should contain no air-entraining agent.
Any admixture used should have been in general commercial use for at least 5 years and have a proven record of success (Fig. 1.10).

Technical requirements include:

(a) Main effect of admixture, water reducing, retarding, or accelerating
(b) Any additional influences admixture may have
(c) Physical properties of the material
(d) Concentration of active ingredient
(e) Presence of any potentially detrimental substances such as chlorides, sulfates, etc.
(f) pH
(g) Potential occupational hazards for users
(h) Conditions for storage and shelf life
(i) Instructions for preparation of admixture and procedures for introducing it into the concrete mix

Dosage rates for chemical admixtures should be in accordance with manufacturer's recommendations; however, run relevant tests to confirm effects with the job materials in the field. Since chemical admixtures may influence the maintenance of

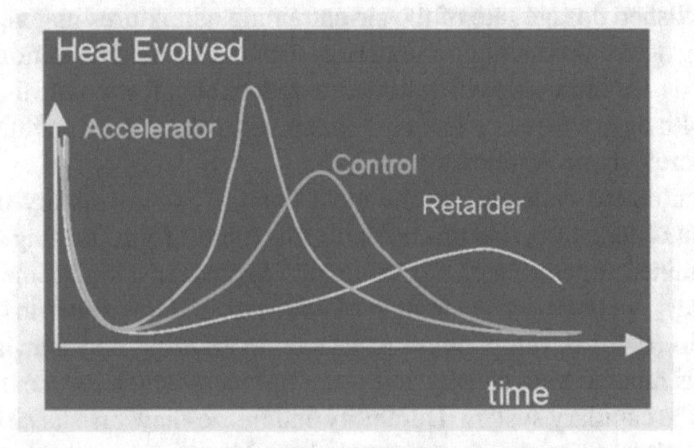

Fig. 1.10 Schematic effect of heat generated by various chemical admixtures

the required entrained air in concrete, the compatibility of specific chemical and air-entraining admixtures should be evaluated prior to production.

Ensure reliable procedures are established for accurate batching of the admixture, especially with chemical admixtures which may have dosages below 0.1% by weight of cement. (Overdoses can easily occur with disastrous results.)

Be aware of the compatibility effects the admixture can have on other concrete properties as most affect several concrete properties.

High-Range Water Reducers, Types F and G: Poly-carboxylate-based, conform to ASTM C494 Type F, to achieve consistent workability with a target slump of 5–8 inches. These are typically used where self-consolidating concrete is required or dictated by placement conditions (congestion of reinforcing steel).

Type C Accelerating Admixture: Liquid-based admixtures, designed to accelerate the strength of concrete under freezing weather.

ASTM C494 also include Type S, specific performance admixture: Some of the specific performance admixtures commercially available are as follows.

Shrinkage-Reducing Admixture (SRA) SRAs have proven effective in reducing plastic and autogenous shrinkage. These chemical admixtures reduce the surface tension of the pore solution minimizing capillary stresses and drying shrinkage. The admixture remains in the pore system after the concrete has hardened and continues to reduce surface tension. The use of the SRA has a slight retarding effect on the rate of cement hydration and may extend the setting time. According to the available data, addition of SRA in concrete reduces the shrinkage up to 40% at 28 days. The recommended dosage rates vary (0.75–1.50 gallon/cy of the concrete mixture) for effectiveness on length change and are added in liquid form. When the SRA is added, the same amount of water is taken out of each mix.

Corrosion Inhibitors Corrosion inhibitors, as the name implies, are designed to inhibit the corrosion of steel in reinforced concrete. These are nitrate-based admixtures that have a high affinity to steel and displace chloride ions from the metal surface to concrete and reduce chloride-induced corrosion. Such an admixture provides an effective means for extending the service life of concrete exposed to chlorides. The long-term performance studies of these admixtures, being relative new, are limited (Fig. 1.11).

1.8 Corrosion Resisting Reinforcing (CRR) Steel

High-chromium reinforcement can extend the design life of a bridge deck approximately five times longer when compared to epoxy-coated reinforcement (ECR). Bridge decks constructed with ECR generally need a protective overlay applied after approximately 40 years. This suggests that using a steel bar with at least 9.2% chromium may not require a protective overlay for up to 100 years after construction, decreasing the whole life costs considerably [2].

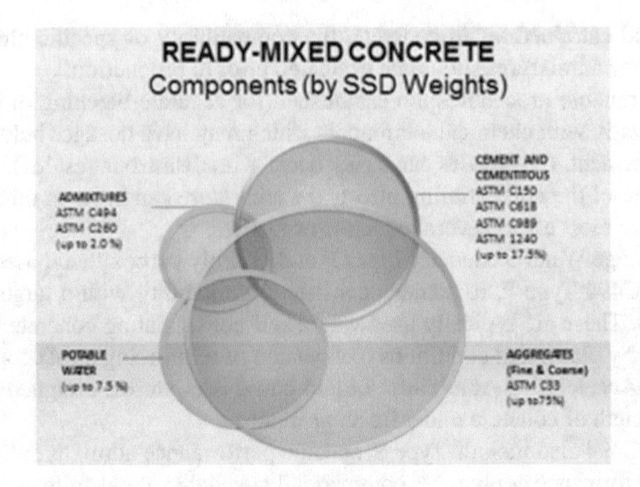

Fig. 1.11 Schematic of ready-mixed concrete components by mass

Fig. 1.12 Photograph of embedded low-carbon, high-chromium steel bars. (a) As-received bars, (b) pristine bars [3]

Solid stainless steel rebar can also meet the design life requirement of 100 years, but the higher price is hard to justify especially across large cross sections/structures. A solid stainless steel would require less concrete cover and reduce concrete cost and weight. A lower price option in the form of low-carbon chromium steel alloy is now available.

ASTM A1035 specification covers the requirements deformed, low-carbon (0.15–0.3%), chromium (2.0–10.9%) alloy steel for concrete reinforcement. Several State DOTs do allow steel with a minimum 9.2% chromium as suitable for extended service life for bridge structural elements including deck reinforcement. Examples of low-carbon, high-chromium steel bars as received and with mill scale removed (pristine condition) are shown in Fig. 1.12 [3].

Stainless Steel Stainless steels are generally grouped into martensitic, austenitic, ferritic, and duplex stainless steels. Each group has differing properties based on their combination of alloying elements that are combined with iron (Fe) and chromium (Cr). These elements can include but are not limited to nickel (Ni), molybdenum (Mo), nitrogen (N), silicone (Si), copper (Cu), and carbon (C). The general properties of each group of stainless steel are detailed below [4]:

Martensitic steels can be high- or low-carbon steels that are built around the Type 410 composition of iron (12% Cr and up to 1.2% C). To further improve the properties of this steel alloy, it is tempered. If martensitic steel is not tempered, it becomes brittle and therefore has limited applications.

Austenitic steels contain high levels of chromium and nickel and low levels of carbon; they are commonly known for their formability and resistance to corrosion and are the most widely used form of stainless steel [9]. The level of nickel alloyed promotes the formation of austenite rather than ferrite stainless steels. An austenitic stainless steel generally shows a minimal attraction to a magnet but a high cost due to the added nickel [5].

Ferritic stainless steels have high-chromium content and low-carbon content; they are known for their good ductility and resistance to corrosion and stress corrosion cracking. Ferritic stainless steels are much lower in cost than austenitic due to the lower levels of nickel and also show a much stronger magnetic attraction [5].

Duplex are a combination of ferritic and austenitic stainless steel and will exhibit an intermediate response to a magnet. Duplex stainless steels are about twice as strong as regular austenitic or ferritic stainless steels. Duplex steels have a lower nickel and molybdenum than their austenitic counterparts of similar corrosion resistance and therefore can also be lower in cost. It may also be possible to reduce the section size due to the increased strength of the section [4].

Stainless steels are given specific grades depending on their composition and are selected based on their corrosion resistance, cost, availability, and mechanical properties. The acceptable grades of steel that can be used for CRR are detailed in Table 1.21. The corrosion resistance depends mostly on the composition of the stainless steel. For chloride pitting and crevice corrosion resistance, their chromium (Cr), molybdenum (Mo), and nitrogen (N) contents are most important. The composition of each acceptable grade of CRR steel is detailed in Table 1.21.

Low-Carbon ChromX as CRR MMFX manufactures low-carbon chromium steel reinforcing deformed bars ChromX 9000, conforming to ASTM A 1035/1035M CS, in two grades, Grades 100 and 120 (100 ksi and 120 ksi yield strength, respectively). A comparison of stainless steel and ChromX 9000 is presented in Table 1.22.

ChromX 9000 steel, which provides protection for severe corrosive environment caused by seawater, can meet 100-year service life requirements for infrastructure projects. This is due to the high-chromium content which ensures a chromium surface coats the steel protecting the reinforcement from any chemicals preventing corrosion.

Table 1.21 Acceptable stainless steel grades and their properties

Stainless steel grade	CRR class	Steel type	Chemical composition (weight %) [11–15] Cr	C	Ni	Mo	N	ASTM grade
S24100	Class I	Austenitic	16.5–19.5	<0.15	0.5–2.5	–	0.2–0.45	XM-28
S32101	Class II	Duplex	21–22	<0.04	1.35–1.7	0.1–0.8	0.2–0.25	–
S24000	Class III	Austenitic	17–19	<0.08	2.25–3.75	–	0.2–0.4	XM-29
S30400	Class III	Austenitic	18–20	<0.07	8–10.5	–	–	304
S31603	Class III	Austenitic	16–18	<0.03	10–14	2–3	<0.1	316L
S31653	Class III	Austenitic	16–18	<0.03	10–14	2–3	0.1–0.16	316LN
S31803	Class III	Duplex	21–23	<0.3	4.5–6.5	2.5–3.5	0.08–0.2	A815
S32304	Class III	Duplex	22–24	<0.03	3.5–5.5	0.1–0.6	0.05–0.2	–

Table 1.22 Comparison of stainless and ChromeX9000 steel properties and strength requirements

Reinforcing type	Symbol	Standard	Carbon content %	Chrome content %	Min yield strength ksi	Min. tensile strength, ksi
Low-carbon chrome	ChromX 9000	ASTM A 1035 CS	0.15	8–10.9	100	150
Stainless steel	S24100	AASHTO MP-18M	<0.15	17–19	60	90
	S24000	AASHTO MP-18M	<0.08	16.5–19	60	90

Many State DOTs allow low-carbon high-chromium, ChromX 9000, steel as acceptable materials to be used as CRR for bridges. Examples of Class I low-carbon, high-chromium steel reinforcement, meeting ASTM A1035 CS include VDOT Bridges (e.g., ChromX 9100, Grade 75, used as replacement for the stainless steel, for the construction of VDOT Lesner Bridge over Lynn Haven Inlet, Virginia Beach).

1.9 Conclusions and Recommendations

Concrete is a heterogeneous material, composed of an active ingredient in the form of cement, which acts as a binder upon hydrating, encapsulating the aggregates, and to form a stable material.

Various types of Portland cements are available with varying chemical composition, heat of hydration, and strength requirements.

Supplementary cementitious materials are available to improve hardened concrete properties for applications in aggressive environments.

Fine and coarse aggregates should conform with the applicable industry standards. Use of well-graded aggregates with durable properties to resist wreathing is recommended. Marginal aggregates may be used provided they are qualified by testing.

Air-entraining admixtures aid in workability of concrete and provide a system of air voids in concrete to resist freezing and thawing cycles. Perform ASTM C666 test to evaluate the resistance of concrete mixture for freeze-thaw properties, prior to use.

Chemical admixtures provide increased benefits for concrete in facilitating placement and finishing and promoting in-service durability. The actual dosage of the admixture to achieve the specified benefit should be established by testing prior to use.

Corrosion-resistant reinforcing steel should be considered for design of reinforced concrete structures where service life of up to 100 years is specified (Refer to Chapter 11).

References

1. Panchalan RK, Ramakrishnan V. Validity of 0.45 power chart in obtaining the optimized aggregate gradation for improving the strength aspects of high-performance concrete, ACI Special Publication SP-243, pp. 99–108
2. Weyers R, Sprinkel M, Brown M. (2006) Summary report on the performance of epoxy-coated reinforcing steel in Virginia. Virginia Transport Research Council, pp. 06–r29.
3. Stephen S, Moruza A. (2009) Field Comparison of the Installation and Cost of Placement of Epoxy-Coated and MMFX 2 Steel Deck Reinforcement: Establishing a Baseline for Future Deck Monitoring. Virginia Transport Research Council, pp. 09–r9.
4. Stainless Steels Classifications [Internet]. Engineeringtoolbox.com. 2018 [cited 3 August 2018]. Available from: https://www.engineeringtoolbox.com/stainless-steel-classifications-d_368.html.
5. Sharp S, Lundy L, Nair H, Moen C, Johnson J, Sarver B (2017) Acceptance procedures for new and quality control procedures for existing types of corrosion reinforcement steel. Virginia Center for Transportation and Research, pp. 11–r21.

Chapter 2
Concrete Mixture Design

2.1 Concrete Trial Batches

The proposed mix design should be evaluated from laboratory or field tests in accordance with ACI 211.1. Trial batches should conform to the materials, proportions, slump, mixing and placing equipment, and procedures to be used in the actual work. In the laboratory, a minimum size of 2.0 cu. ft. mixer is recommended for obtaining test samples and making test cylinders for compressive strength.

The materials for concrete should be proportioned to produce a concrete capable of being deposited to obtain maximum density with maximum smoothness of surface.

Test cylinders should be molded from the trial batch containing the maximum water content indicated by the mix design. Each test cylinder should be 6 × 12 inches or 4 × 8 inches; the latter size is being permitted on some projects. Test cylinders at 7 and 28 days. 56-day strength (or later) can be specified where early strengths are not required. A minimum of two cylinders should be tested at each age. Spare cylinders should be cast to be tested at early ages (for early removal of forms) or later ages (e.g., 56 days for concrete containing fly ash) to establish the compressive strength at various ages, prior to production.

The placing of concrete specified by compressive strength should not begin until the mix design has been qualified in accordance with the test criteria given in Tables 2.1, 2.2, 2.3, and 2.4. If the sources of materials or established procedures change, new trial batches should be tested and qualified. For large projects, where an on-site concrete batch plant is required, the laboratory concrete trial mixtures should be further verified in the field, prior to production.

ACI 318: Building Code Requirements for Structural Concrete Durability
ACI 318 requires the licensed design professional engineer to assign exposure class of concrete based on the severity of the anticipated exposure of structural concrete members based on exposure classes.

N. Hasan, *Durability and Sustainability of Concrete*,
https://doi.org/10.1007/978-3-030-51573-7_2

Table 2.1 Concrete classes exposed to freezing and thawing[a]

Class	Freezing and thawing exposure	Air content	Minimum strength f′c (psi)	Maximum water/cementitious material ratio
F0	Not applicable	N/A	2500	N/A
F1	Moderate	Table 4.4.1	4500	0.45
F2	Severe	Table 4.4.1	4500	0.45
F3	Very severe	Table 4.4.1	4500	0.45

[a]Refer to ACI 318 Building Code, Table 4.3.1. For Classes F1, F2, and F3, concrete must be air entrained per Table 4.4.1

Table 2.2 Concrete classes exposed to sulfates[a]

Class	Sulfate exposure	Cementitious material[b]	Minimum strength f′c (psi)	Maximum water/cementitious material ratio
S0	Not applicable	No restriction	2500	N/A
S1	Moderate (150<SO_4 <1500 ppm)	Use Type II cement or blended cement	4000	0.45
S2	Severe (150<SO_4 <10,000 ppm)	Use Type V cement or blended cement	4500	0.45
S3	Very severe (SO_4 >10,000 ppm)	Use Type V cement and SCMs	4500	0.45

[a]Refer to ACI 318 Building Code, Table 4.3.1
[b]Alternative combinations of cementitious materials, including natural pozzolans, silica fume, and slag, may be used to improve sulfate resistance of concrete, based on testing

Table 2.3 Concrete classes exposed to low permeability and protection against corrosion[a]

Permeability/ corrosion class	Exposure	Maximum water-soluble chloride ion in concrete, by weight of cement	Minimum strength f′c (psi)	Maximum water/ cementitious material ratio
P0	Not applicable	No restriction	2500	N/A
P1	Low permeability	None	4000	0.50
C0	N/A	1.00	2500	N/A
C1	Severe	0.30	2500	N/A
C2	Very severe	0.15	5000	0.40

[a]Refer to ACI 318 Building Code, Table 4.3.1

Table 2.4 Total air content for concrete exposed to cycles of freezing and thawing[a]

	Air content, percent	
NMSA, in.	Exposure Class F1	Exposure Classes F2 and F3
3/8	7	7.5
1/2	5.5	7
3/4	5	6
1	4.5	6
1–1/2	4.5	5.5
2	4	5
3	3.5	4.5

[a]Refer to ACI 318 Building Code, Table 4.4.1

ACI 318 requirements for concrete by exposure class, including minimum strength, w/cm ratio, limits on sulfates, or chlorides by exposure class, are presented in Tables 2.1, 2.2, and 2.3. The mix proportions should be adjusted during the progress of work whenever need for such adjustment is indicated by results of testing.

Exposure Category F applies to exterior concrete exposed to moisture cycles of freezing and thawing (Table 2.1).

Exposure Class S applies to concrete exposed to water containing deleterious amounts of water sulfate irons (Table 2.2).

Exposure P applies to concrete in contact with water requiring low permeability.

Exposure C applies to concrete exposed to conditions that require additional protection against corrosion of reinforcement (Table 2.3).

When a structural concrete is exposed to more than one exposure class, the most restrictive requirements are applicable. For example, a structural concrete member, assigned Exposure Class F3 and Exposure Class C2, would require concrete to comply with the minimum f′c of 5000 psi and a maximum *w/cm* ratio of 0.4, since the requirement for corrosion protection is more restrictive than the requirement for freezing and thawing.

For a given project location, the exposure class for concrete subject to freezing and thawing may be established in accordance with US Weathering Region Map, Fig. 2.1 of ASTM C33 [3], which subdivides the regions into negligible weathering (F1), moderate weathering (F2), and severe weathering (F3).

ACI 318 imposes additional requirements for air entrainment for concrete exposed to freezing and thawing cycles, as shown in Table 2.4. For f′c greater than 5000 psi, reduction of air content by 1.0 percent of the listed values is permitted.

ACI Additional Requirements for Concrete Subject to Class F3 Exceptions to the target air content values for Exposure Classes F1, F2, and F3 are permitted by ACI 318. For concrete with f′c greater than 5000 psi, Paragraph 4.4.1 of ACI 318 permits one percent lower air content than the target values for each exposure class.

Fig. 2.1 San Roque Multipurpose Project aerial view upon completion

Table 2.5 Requirements for establishing suitability of cementitious material combinations exposed to water-soluble sulfates[a]

Exposure class	Maximum expansion when tested using ASTM C1012		
	At 6 months	At 12 months	At 18 months
S1	0.10 percent		
S2		0.10 percent	
S3			0.10 percent

[a]ACI 318 Table 4.5.1

For a given project location, the exposure class for concrete subject to freezing and thawing may be established in accordance with US Weathering Region Map, Fig. 2.1 of ASTM C33, that subdivides the regions into negligible weathering (F1), moderate weathering (F2), and severe weathering (F3).

Exceptions to the air content for Exposure Classes F1 and F2 may be reviewed and modified by on a case-by-case basis, which are applicable to concrete mixes with compressive strength of 5000 psi (or greater), low water-to-cementitious material ratio, concrete with dry density of 140 lb per cu ft, and exposure conditions for concrete such that there is little or no in contact with moisture during freezing and thawing cycles. Such requirements including exposure class for freezing and thawing should be defined in the Contract Documents.

ACI 318 recommends the evaluation of sulfate resistance of concrete mixtures using alternative combinations of SCM in accordance with ASTM C1012. The expansion criterion for each Exposure Class S1, S2, and S3 is listed in Table 2.5.

2.2 Concrete Design Compressive Strength

The design compressive strength (f′c) for all concrete classes, including concrete containing fly ash, is established at the age of 28 days, as indicated in design drawings and specifications unless otherwise directed by the design professional. The design compressive strength for concrete, made with fly ash or with SCM, may be established at the age of 56 days in consultation with the design professional and construction schedule.

ACI 211.1 methods for selecting and adjusting proportions for concrete provides two methods for selecting and adjusting proportions for normal weight concrete: the estimated weight and absolute volume methods.

ACI 318 requires that the required average compressive strength (f′cr) of concrete shall be based on the standard deviation(s) of the test data available from an existing production facility or field tests as presented in Tables 2.6, 2.7, and 2.8.

ACI 318 requires a modification to the sample standard deviation when less than 30 tests are available, as shown in Table 2.7.

When a production facility has no field strength records for calculation of standard deviation(s), ACI 318 requires that the required average compressive strength (f′cr) shall be determined in accordance with requirement given in Table 2.8.

Table 2.6 Required average compressive strength when data are available to establish a sample standard deviation[a]

Specified compressive strength, psi	Required average compressive strength, psi
f′c <5000	Use the larger value computed from *f′cr = f′c+1.34s, or* *f′cr = f′c+2.34s-500*
f′c >5000	Use the larger value computed from *f′cr = f′c+1.34s, or* *f′cr = 0.9f′c+2.34s*

f′c Design compressive strength, *f′cr* Required compressive strength
[a]Refer to ACI 318 Building Code, Table 5.3.2.1, where

Table 2.7 Modification factor for sample standard deviation when less than 30 tests are available[a]

Number of tests	Modification factor for standard deviation
Less than 15	Use Table 5 above
15	1.08
20	1.03
25	1.03
30	1.00

[a]Refer to ACI 318 Building Code, Table 5.3.1.2

Table 2.8 Required average compressive strength when data are not available to establish a sample standard deviation[a]

Specified compressive strength, psi	Required average compressive strength, psi
3000<f′c <5000	f′cr = f′c+1200
f′c >5000	f′cr = 1.1f′c+700

[a]Refer to ACI 318 Building Code, Table 5.3.2.2

2.3 Requirements for Massive Concrete Mixtures

Massive Concrete: Structures with large dimensions that require special measures be taken with temperature and with the generation of heat and attended volume change to minimize cracking. ACI 207 do not require a specific member thickness for mass concrete. Mass concrete may be defined in terms of the least dimension of the section to be placed. Concrete volume may be considered massive if the least dimension to be placed equals or exceeds 4 feet. The requirements for mass concrete mixture should include, but are not limited to, the following principles:

1. Include fly ash (ASTM C618, Type F) and/or SCM up to 50% maximum by weight of cement replacement. The final mix proportions should be established by trial mixes. Use minimum weight of cement where possible.
2. Type II (MH) cement with moderate heat of hydration.
3. Largest maximum size coarse aggregate, consistent with the reinforcing spacing and concrete cover requirements.
4. Consider thermal analyses, including a thermal control plan (TPC) of the concrete placement to determine maximum permissible concrete temperature during curing.
5. Consider using internal cooling systems.
6. The placement schedule for mass concrete should allow for alternate placing sequence to control volume change and shrinkage cracking. The time interval between adjoining placements should be specified.
7. The curing of mass concrete should be performed by extending the wet curing for a minimum of 7 days.

Additional information for mass concrete placement including TCP are provided in Chap. 5.

Case Study: Paerdegat Basin Bridge Mass Concrete, Brooklyn, NY

Class HP concrete, with a compressive strength of 3000 psi at 28 days, was selected for mass concrete for the concrete footings and superstructure of new Shore (Belt) Parkway Bridge over Paerdegat Basin, Brooklyn, New York. The concrete mixture proportions are presented in Table 2.9. In addition, a thermal control plan (TCP) was implemented (refer to Chap. 5 for details):

1. Center of concrete temperature was limited to 160F maximum during curing.
2. The temperature differential between the center and surfaces of each mass placement was limited to 35F, in accordance with NYC DOT Specifications 555.02.

Table 2.9 Class HP concrete mixture proportions (3000 psi at 28 days)

Ingredients	Weight, kg	Specific gravity	Remarks
Water	162	1.0	W/CM = 0.40
Cement	301	3.15	Holcim (College Point), NY
Sand	636	2.62	Roanoke (FM 2.85)
Stone	1089	2.81	Tilcon
Silica fume	24	2.26	Euclid
Fly ash	80	2.2	Pro Ash-STI
Air	600 ml		BASF-MBVR (5–8%)
WRA	800 ml		BASF Pozz. 200N
SSRWR	300 ml		BASF-100XR

Table 2.10 Class MP concrete mix proportions based on 56-day design compressive strength of 21 MPA (3000 psi)

	Weight, kg (lbs./yd³ unless noted))	Specific gravity	Remarks
Water	162	1.0	W/CM ratio = 0.40
Cement	178 (300)	3.15	Holcim (College Point), NY
Sand	636	2.62	Roanoke (FM 2.85)
Stone	1089	2.81	Tilcon
Silica fume	0	2.26	Euclid
Fly ash	119 (201)	2.2	Pro Ash-STI
Air	600 ml		BASF-MBVR (5–8%)
WRA	800 ml		BASF Pozz 200N
HRWR	300 ml		BASF-100XR
Adiabatic temperature rise (ATR), F	HP mix	MP mix	Based on heat of hydration of cement, fly ash, and silica fume
1 day	49	32	(Typically fly ash generates one-half the cement heat of hydration)
7 days	84	55	
28 days	96	62	
56 days	99	65	

3. TCP was applicable to sections exceeding 1.22 m (4 ft) thickness, including footings, pier columns, and pier caps. A separate TCP was implemented for each mass placement.
4. All mass placements were provided with insulation, cooling (water) pipes, and sensors for thermal control monitoring. A minimum R2.5 insulation was used for sealing all the exposed surfaces, including protruding reinforcement, during curing.

An alternative Class MP concrete was also considered for mass concrete to lower the maximum temperatures and temperature differences (Table 2.10).

Case Study: San Roque Multipurpose Project, Pangasinan, Philippines

Project Description The 200 m (650 ft) high San Roque Dam on the Lower Agno River on Luzon Island, the Philippines, is an earth and rockfill dam, 12th highest in the world. It was completed on November 19, 2002. The major features of the Project, the 345 MW Power Plant, included 7 km (4.4 miles) of tunnels and subsurface galleries, an 8.5 meter (28 feet) diameter by 1200 meters long (3900 feet) power tunnel and lined penstock, and 5.5 meter (18 feet) diameter by 1400 meters (4600 feet) long mid-level release tunnel. Another major feature was the 100m (328 ft) wide by 400 m (1312 ft) long gated concrete chute spillway, capable of passing the Probable Maximum Flood of 12,800 cms (452,000 cfs).

2.4 Qualification of Materials for Concrete [1]

Aggregate Sources and Reactivity The aggregates for concrete were obtained from the Agno River. A crushing plant was set up on site to process crushed rock for dam rockfill and aggregates for concrete. Samples of fine and coarse aggregates from Agno River, proposed Type II cement, fly ash, and silica fume were collected from the San Roque site, Philippines, and were shipped to the US Army Corps of Engineers, Concrete and Materials Laboratory in Vicksburg, Mississippi, USA, for testing. The purpose was to determine compliance with the specification requirements for potential alkali reactivity of the project aggregates.

Potential Alkali Reactivity The tests for potential alkali-silica reactivity (ASR) of the project aggregates and standard cement were performed in accordance with ASTM C1260, which defines limits for deleterious expansion (refer to Chap. 3 for details on ASR testing). As a result of these tests, a low-alkali cement and fly ash were used in all project concrete.

Selection of Concrete Mixture Proportions An extensive concrete trial mix program was conducted on site to select concrete design mix proportions for a range of strength classes, from 207 Kg/cm^2 (3000 psi) to 695 kg/cm^2 (10,000 psi). Tests were performed for workability, pumpability, and strength. The mixes were developed for coarse aggregate gradations ranging from 9.5 mm (3/8-inch) to 76 mm (3-inch) maximum size, and slumps range from 25 mm (1-inch) to 200 mm (8-inch) to meet the applicable placement requirement and conveying method (bucket/crane, creter crane/conveyors, and pumping methods).

Low-Alkali Cement Due to potential alkali reactivity of the aggregates described above (and in Chap. 3), ASTM C150 Type II low-alkali cement (alkalis: 0.60 percent maximum) was used. Tests on Union cement samples, obtained from the plant and the bulk cement shipments on arrival at site, were routinely tested for compliance with specification. Table 2.11 presents typical test results on the chemical and physical analyses for the San Roque cement.

Heat of Hydration Heat of hydration testing on project cement and project cement at 10 percent replacement with silica fume were performed in accordance with ASTM C186. The testing showed up to 30 percent reduction in heat of hydration for cement-silica fume mixture, as shown in Table 2.12.

False Set False set for project cement, determined in accordance with ASTM C451, was found to be 82 % (vs 50% minimum specified).

Table 2.11 San Roque cement test results

Requirements	ASTM C150 Type II specification	Test results
Chemical analysis		
1. Silicon dioxide, %	20.0 min	21.60
2. Aluminum oxide, %	6.0 max	6.0
3. Ferric oxide, %	6.0 max	4.8
4. Magnesium oxide, %	6.0 max	0.9
5. Sulfur trioxide, %	3.0	2.2
6. Loss on ignition, %	3.0 max	1.1
7. Insoluble residue, %	0.75 max	0.70
8. Tricalcium silicate, %	–	34
8. Dicalcium silicate, %	–	36
10. Tricalcium aluminate, %	8 max	8
11. Tetra-calcium alumino ferrite, %	–	15
12. Calcium oxide, %	–	61.9
Physical analysis		
1. Blaine air permeability test specific surface, m^2/kg	280 min	350
2. Autoclave expansion, %	0.80 max	0.03
3. Compressive strength, psi		
3 days	1450 min	2660
7 days	2470 min	3740
4. Time of setting, Vicat: Method		
Initial set, minutes	45–375	74
Final set, minutes		156

Table 2.12 San Roque cement heat of hydration, cal./g

	3 days	7 days	28 days	Specification
Cement (100%)	59	69	78	70 (7 days) maximum
Cement + 10% silica fume	46	49	N/A	70 (7 days) maximum

Table 2.13 San Roque Class F fly ash test results

Requirements	ASTM C618 Class F specification	Test results
I. Chemical analysis		
1. Silicon dioxide + aluminum oxide + ferric oxide, min %	70.0	83.6
2. Sulfur trioxide, max %	3.0	0.2
3. Moisture content, max %	1.0	0.2
4. Loss on ignition, max %	4.0	2.3
II. Physical analysis		
1. Autoclave expansion or contraction, max %	0.8	0.03
2. Strength activity index		
With Portland cement		
@ 7 days, min % of control	75	76
@ 7 days, min % of control	75	79

Fly Ash Fly ash conforming to ASTM C618, Class F, with limits on loss of ignition (4%), sulfur trioxide content (3%), and moisture content (1%) was specified. Table 2.13 presents typical test results on the fly ash chemical and physical analyses.

Silica Fume Condensed silica fume conforming to ASTM C1240 with the following additional requirements was specified:

Silicon dioxide 90.0 percent minimum
Loss of ignition 6.0 percent maximum
Sulfur trioxide 1.0 percent maximum
Moisture content 3.0 percent

Chemical Admixtures Chemical admixtures conforming to ASTM C494, Types A, D, F, and G, were specified. All admixtures were dispensed in a liquid form.

Aggregates The coarse aggregate and fine aggregate were obtained from a crushing plant on site. The coarse aggregates were generally produced and stockpiled in three sizes, 1–1/2 to 3/8 inch, 3/4 inch to No. 8, and 3/8 inch to No. 16 fractions, conforming to ASTM C33 gradations.

Cement Uniformity Type II cement was routinely tested for physical and chemical properties for compliance with ASTM C150. Samples were tested weekly by a laboratory, Philippines Geoanalytics, Inc. The cement plant was also inspected regularly to monitor quality control. One of the uniformity issues encountered during production of the cement was the rapid slump loss after mixing concrete. It was found that the loss of slump was contributed by lack of control on gypsum. As a result, the gypsum feeder at the plant was upgraded from manual to automatic control.

Concrete Mixtures A total of 21 concrete design mixes were developed for plant structures by class, subclass based on aggregate size, and design compressive strength (28 days or 56 days). Selected concrete mixtures are presented in Table 2.14.

Table 2.14 San Roque selected concrete mixtures and design strength

Mixture design	Structure	Design strength psi
A1-1A	Tunnel lining/powerhouse	4000
A2-11E	Powerhouse/spillway/tunnels	4000
MI-13A	Spillway – mass	4000
M2-02B	Spillway ogee	3000
S1-3/8	LLO tunnel transition	10,000
Si-7A	LLO tunnel transition	10,000
S2-3	Spillway chute	5000
1A	Shotcrete with fibers	5000
2A	Shotcrete without fibers	5000
CG 0010	Rockbolt grout	4000

Table 2.15 Concrete mixtures for powerhouse and spillway (quantities are lbs per cubic yard unless noted)

Mixture		Mixture A 1A	Mixture A2 11E1	Mixture AAA2
Strength		4000 psi	4000 psi	6000 psi
	SG	Weight	Weight	Weight
Cement	3.15	460.3	460.3	621.6
Fly ash	2.5	100.8	189.8	201.6
Sand	2.75	1251.6	1449.8	991.6
#3/8 stone	2.76		235.2	0.0
#67 stone	2.67	1896.7	1253.3	1228.5
#4 stone	2.67		325.9	809.8
Water	1	100.8	160.0	112.6
Ice	1	154.9	96.0	151.2
AEA, oz/cwt	1.1	4	4	5
WRA, oz/cwt	1.1	20	20	4
Total weight		3989	4194	4126
w/cm		0.46	0.39	0.32
Paste		662	810	936
Percentage		0.17	0.19	0.23
Air, %		4.5–7.5	4–6	4.5–7.5
Un Wt	139.6	150.5	154.2	151.7

Concrete mixture proportions for powerhouse (Mixture A 1A) and spillway (Mixture A2 11E1 and Mixture AAA2) concrete are shown in Table 2.15.

Performance of concrete during production, including compressive strength, was monitored. The results are discussed in Chap. 9, Quality Control.

A total of 392,500 cubic meters of concrete was place for the project. The overall concrete placement for the January 2000–April 2002 period is shown in Fig. 2.3.

Figure 2.2 shows a combined gradation analysis for Mixture AAA2 with #67 stone (3/4 inch).

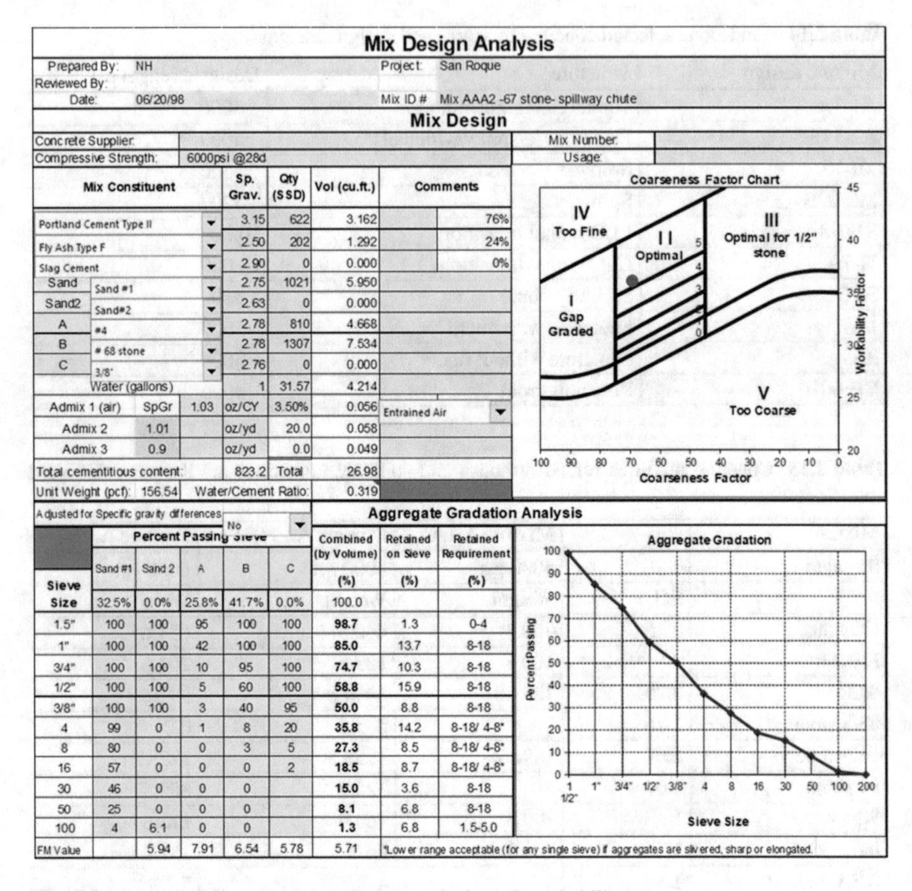

Mix Design Analysis

Prepared By: NH
Reviewed By:
Date: 06/20/98
Project: San Roque
Mix ID #: Mix AAA2 -67 stone- spillway chute

Mix Design

Concrete Supplier:
Compressive Strength: 6000psi @28d
Mix Number:
Usage:

Mix Constituent		Sp. Grav.	Qty (SSD)	Vol (cu.ft.)	Comments
Portland Cement Type II		3.15	622	3.162	76%
Fly Ash Type F		2.50	202	1.292	24%
Slag Cement		2.90	0	0.000	0%
Sand	Sand #1	2.75	1021	5.950	
Sand2	Sand#2	2.63	0	0.000	
A	#4	2.78	810	4.668	
B	# 68 stone	2.78	1307	7.534	
C	3/8"	2.76	0	0.000	
Water (gallons)		1	31.57	4.214	

Admix 1 (air)	SpGr	1.03	oz/CY	3.50%	0.056	Entrained Air
Admix 2	1.01		oz/yd	20.0	0.058	
Admix 3	0.9		oz/yd	0.0	0.049	
Total cementitious content:			823.2	Total	26.98	
Unit Weight (pcf):	156.54	Water/Cement Ratio:			0.319	

Adjusted for Specific gravity differences: No

Aggregate Gradation Analysis

Sieve Size	Sand #1	Sand 2	A	B	C	Combined (by Volume) (%)	Retained on Sieve (%)	Retained Requirement (%)
	32.5%	0.0%	25.8%	41.7%	0.0%	100.0		
1.5"	100	100	95	100	100	**98.7**	1.3	0-4
1"	100	100	42	100	100	**85.0**	13.7	8-18
3/4"	100	100	10	95	100	**74.7**	10.3	8-18
1/2"	100	100	5	60	100	**58.8**	15.9	8-18
3/8"	100	100	3	40	95	**50.0**	8.8	8-18
4	99	0	1	8	20	**35.8**	14.2	8-18/ 4-8*
8	80	0	0	3	5	**27.3**	8.5	8-18/ 4-8*
16	57	0	0	0	2	**18.5**	8.7	8-18/ 4-8*
30	46	0	0	0		**15.0**	3.6	8-18
50	25	0	0	0		**8.1**	6.8	8-18
100	4	6.1	0	0		**1.3**	6.8	1.5-5.0
FM Value		5.94	7.91	6.54	5.78	5.71		

*Lower range acceptable (for any single sieve) if aggregates are silvered, sharp or elongated.

Fig. 2.2 Concrete Mixture AAA-2 design analysis and workability

Case Study: Sidney A. Murray Jr. Hydroelectric Station, Black Hawk, LA (1985–1990) [2]

This 192 MW Sidney A. Murray Jr. Hydroelectric Station is the nation's largest low-head power plant. The plant consists of eight bulb turbine generators with 27-foot-diameter runners (the world's largest to date), with operating heads ranging from 8 to 20 feet between the Mississippi River and the Old River control outflow channel (Fig. 2.4).

The steel structure for the power plant was built in a shipyard and towed to the site, thereby allowing simultaneous construction of the powerhouse and foundation and reducing the time and money required by conventional construction methods (Fig. 2.5). The steel structure was then filled with self-consolidating concrete. The hydroelectric complex is integrated into the existing Corps of Engineers' Old River control system, passing up to 180,000 cfs of flow from the Mississippi to the Atchafalaya Basin. In addition, the powerhouse and abutment dams form a part of the Corps of Engineer Mississippi River mainline levee system. Located in

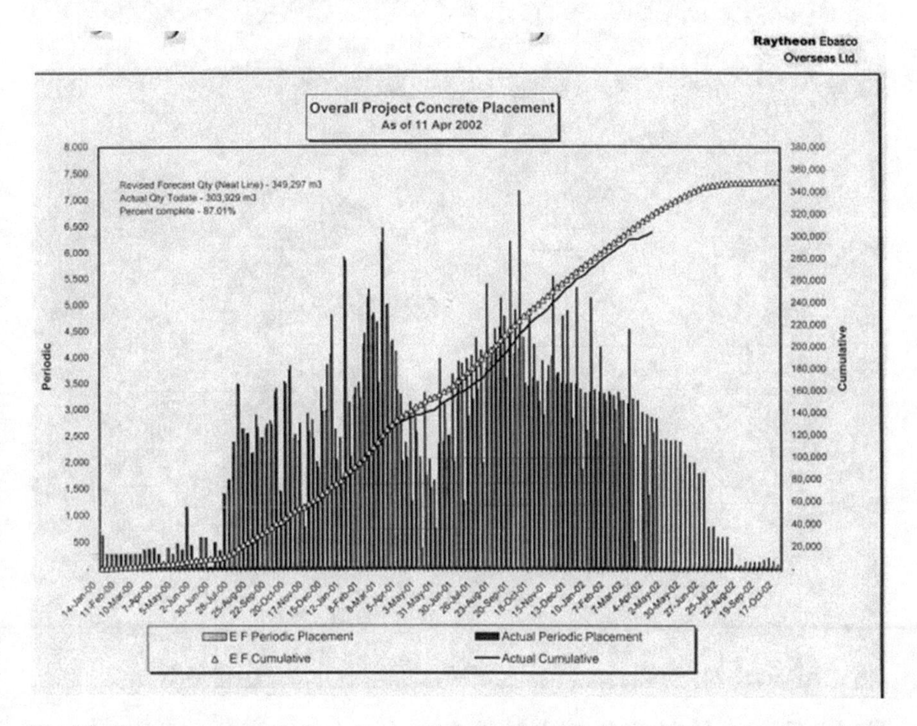

Fig. 2.3 Concrete actual periodic and cumulative placement for San Roque Project over the 2000–2002 period (cubic meters)

Louisiana, the Mississippi River main line levee system was breached and modified for the project. The project involved, in part, a 5000-foot intake channel from the Mississippi and a 9000-foot outlet channel to the Old River overflow channel, 100-ft-high concrete abutment dams, relocation of 5000 feet of levee, and construction of 4000 feet of new levees. These canals and levees required over 2.5 million cubic yards of dry excavation and 15 million cubic yards dredged excavation.

When the channel was complete, more than 600,000 cubic yards of riprap were placed along the slopes.

The design required a composite steel/concrete performance of the structure, requiring concrete to fill each compartment completely without voids. To minimize thermal cracking in concrete from heat of hydration a maximum temperature of 118F on concrete during setting and hardening was specified. The design compressive strength of concrete at 28 or 90 days for concrete containing fly ash was established at 3000 psi.

A comprehensive field model testing program, involving a total of twelve concrete mixtures, six with Melment HRWRA and six without HRWRA, was executed to define the mix design and concrete placement procedure. It was recognized from the early design stage that the concrete design mixture must incorporate both a Melment-based HRWRA and Class F fly ash to meet the requirements of flowability and to reduce heat of hydration [1].

The selected concrete mixture proportions for the super-plasticized concrete (SCC) are shown in Table 2.16.

Fig. 2.4 Sidney A. Murray Jr. Hydroelectric Station

Fig. 2.5 Steel power plant structure being moved to the flooded site for in-the-wet placement

Table 2.16 Concrete mixture proportions for SCC concrete (quantities are lb. per cubic yard unless noted)

Design mixture	SG	3A.2S	3A.7S	3A.8S
Type II cement	3.15	351	320	320
Fly ash	2.2	83	193	202
Sand SSD	2.61	1572	1219	1502
3/4-inch aggregate, SSD	2.73	1608	1811	1528
Water	1	234	213	213
AEA, oz/cy	1.1	5	8	4
WRDA, oz.	1.1	24	23	23
HRWRA, oz.	1.1	70	101	101
				0
Total, lbs/cy		3947	3888	3893
w/c		0.54	0.42	0.41
Slump, inches		9	9.5	9.75
Air, %		6	4.5	5
Initial setting time (hr:min)			9:24	8:09
Compressive strength, psi				
7 days		3100	2636.0	2080
28 days		4560	3813.0	3426
90 days			4381.0	4706

Tests were performed on concrete at the batch plant to verify compliance with the specification requirements. The SCC was tested for temperature, flowability, and cylinders for strength tests every 150 yd^3 delivered. Due to good-quality control, deviations from the specification were kept to a minimum.

Combination of lignin-based water reducer and fly ash increases the initial setting time of concrete mixture. Mixture 3A.2S, with 18% Class F fly ash and w/cm ratio of 0.54, had the lowest slump loss of 1 inch in 20 minutes after batching. Mixture 3A.2S design analysis including gradation, coarseness factor, and workability is shown in Fig. 2.6. Mixture 3A.8S with HRWRA and 38% fly ash had a w/cm ratio of 0.41 and was used for massive portions of the PPS concrete. Mixture 3A.8S design analysis including gradation, coarseness factor, and workability is shown in Fig. 2.7.

Case Study: Rainbow Hydroelectric Redevelopment, Great Falls, MT (2009–2012) (Figs. 2.8 and 2.9)

The Rainbow Redevelopment Project is located on the Missouri River about 6 miles northeast of the City of Great Falls, Montana. The original 36 MW powerhouse and ancillary structures were replaced with a new 62 MW powerhouse.

Project Description This work included design for the new reservoir intake, the power canal, the new penstock intake, forebay, surge overflow structure, penstock, and the powerhouse. Specifically, it included:

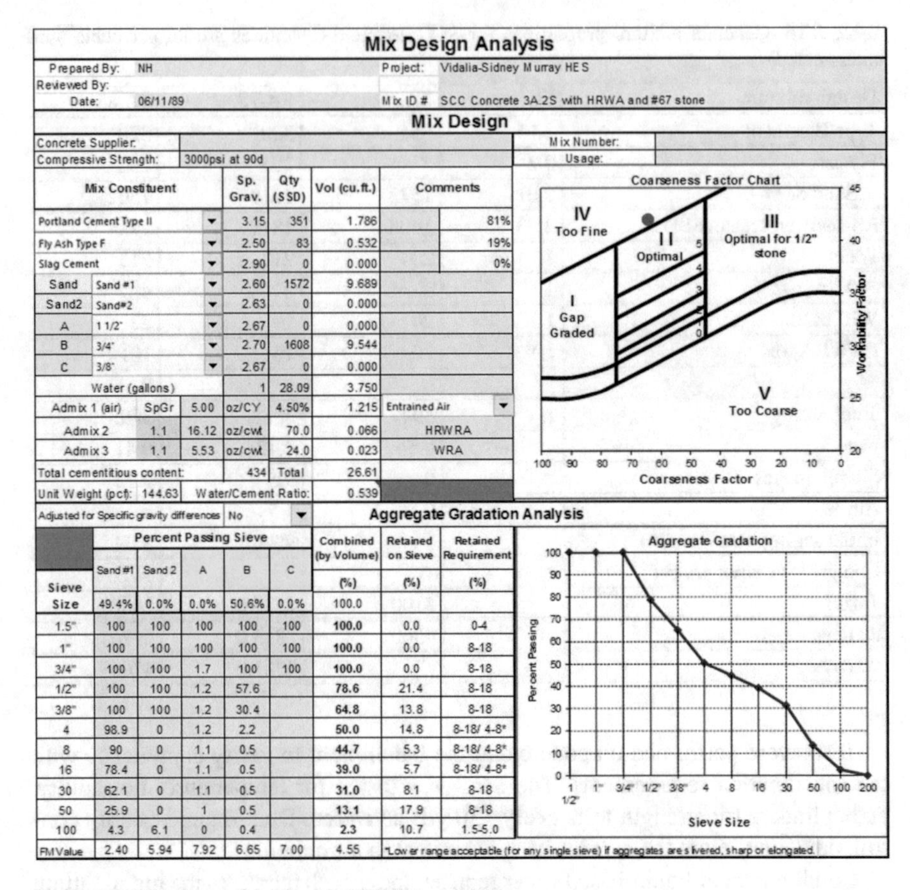

Mix Design Analysis

Prepared By:	NH	Project:	Vidalia-Sidney Murray HES
Reviewed By:			
Date:	06/11/89	Mix ID #	SCC Concrete 3A.2S with HRWA and #67 stone

Mix Design

Concrete Supplier:		Mix Number:	
Compressive Strength:	3000psi at 90d	Usage:	

Mix Constituent		Sp. Grav.	Qty (SSD)	Vol (cu.ft.)	Comments	
Portland Cement Type II		3.15	351	1.786		81%
Fly Ash Type F		2.50	83	0.532		19%
Slag Cement		2.90	0	0.000		0%
Sand	Sand #1	2.60	1572	9.689		
Sand2	Sand#2	2.63	0	0.000		
A	1 1/2"	2.67	0	0.000		
B	3/4"	2.70	1608	9.544		
C	3/8"	2.67	0	0.000		
Water (gallons)		1	28.09	3.750		

Admix 1 (air)	SpGr	5.00	oz/CY	4.50%	1.215	Entrained Air
Admix 2	1.1	16.12	oz/cwt	70.0	0.066	HRWRA
Admix 3	1.1	5.53	oz/cwt	24.0	0.023	WRA
Total cementitious content:			434	Total	26.61	
Unit Weight (pcf):	144.63	Water/Cement Ratio:			0.539	

Adjusted for Specific gravity differences: No

Aggregate Gradation Analysis

Sieve Size	Percent Passing Sieve Sand #1	Sand 2	A	B	C	Combined (by Volume) (%)	Retained on Sieve (%)	Retained Requirement (%)
	49.4%	0.0%	0.0%	50.6%	0.0%	100.0		
1.5"	100	100	100	100	100	100.0	0.0	0-4
1"	100	100	100	100	100	100.0	0.0	8-18
3/4"	100	100	1.7	100	100	100.0	0.0	8-18
1/2"	100	100	1.2	57.6		78.6	21.4	8-18
3/8"	100	100	1.2	30.4		64.8	13.8	8-18
4	98.9	0	1.2	2.2		50.0	14.8	8-18/ 4-8*
8	90	0	1.1	0.5		44.7	5.3	8-18/ 4-8*
16	78.4	0	1.1	0.5		39.0	5.7	8-18/ 4-8*
30	62.1	0	1.1	0.5		31.0	8.1	8-18
50	25.9	0	1	0.5		13.1	17.9	8-18
100	4.3	6.1	0	0.4		2.3	10.7	1.5-5.0
FM Value	2.40	5.94	7.92	6.65	7.00	4.55		

*Lower range acceptable (for any single sieve) if aggregates are slivered, sharp or elongated.

Fig. 2.6 SCC Mixture 3A.2S with HRWRA and #67 stone

A permanent bridge for the access road to the new powerhouse.

A new trashrack structure with trashracks sized for a maximum average flow velocity of 2.5 ft. per second (fps) with a mechanical trashrake and a new power canal intake structure with wheel gates.

A new reservoir intake structure with trashracks, intake gates, and a mechanical trashrake sized for a maximum gross flow velocity of 2.5 fps. The intake is 69.5 feet wide by 26.5 feet high. The trashrack is 105 feet wide by 36 feet high.

An open channel flume type power canal 2375 feet long, 40 feet wide, and 27 feet high.

A penstock forebay and an intake structure with trashrack and wheel gates that lead into a 25-foot-diameter, 380-foot-long steel penstock with compound bends. The penstock intake contains two bays, each 14 feet wide.

A new powerhouse containing a single vertical shaft Kaplan turbine and generator set with a nominal rated capacity of 62 MW.

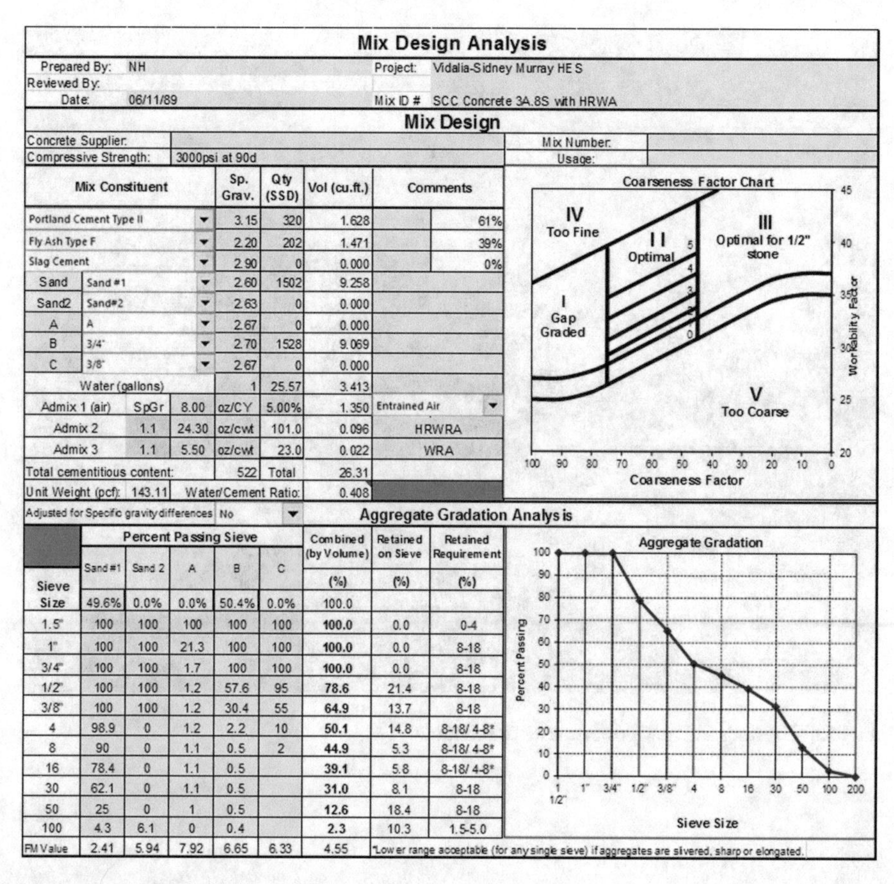

Mix Design Analysis

Prepared By:	NH	Project:	Vidalia-Sidney Murray HES
Reviewed By:			
Date:	06/11/89	Mix ID #	SCC Concrete 3A.8S with HRWA

Mix Design

Concrete Supplier:		Mix Number:	
Compressive Strength:	3000psi at 90d	Usage:	

Mix Constituent		Sp. Grav.	Qty (SSD)	Vol (cu.ft.)	Comments
Portland Cement Type II		3.15	320	1.628	61%
Fly Ash Type F		2.20	202	1.471	39%
Slag Cement		2.90	0	0.000	0%
Sand	Sand #1	2.60	1502	9.258	
Sand2	Sand#2	2.63	0	0.000	
A	A	2.67	0	0.000	
B	3/4"	2.70	1528	9.069	
C	3/8"	2.67	0	0.000	
Water (gallons)		1	25.57	3.413	

Admix 1 (air)	SpGr	8.00	oz/CY	5.00%	1.350	Entrained Air
Admix 2	1.1	24.30	oz/cwt	101.0	0.096	HRWRA
Admix 3	1.1	5.50	oz/cwt	23.0	0.022	WRA
Total cementitious content:			522	Total	26.31	
Unit Weight (pcf):	143.11	Water/Cement Ratio:			0.408	

Adjusted for Specific gravity differences: No

Aggregate Gradation Analysis

Sieve Size	Sand #1	Sand 2	A	B	C	Combined (by Volume) (%)	Retained on Sieve (%)	Retained Requirement (%)
	49.6%	0.0%	0.0%	50.4%	0.0%	100.0		
1.5"	100	100	100	100	100	**100.0**	0.0	0-4
1"	100	100	21.3	100	100	**100.0**	0.0	8-18
3/4"	100	100	1.7	100	100	**100.0**	0.0	8-18
1/2"	100	100	1.2	57.6	95	**78.6**	21.4	8-18
3/8"	100	100	1.2	30.4	55	**64.9**	13.7	8-18
4	98.9	0	1.2	2.2	10	**50.1**	14.8	8-18/ 4-8*
8	90	0	1.1	0.5	2	**44.9**	5.3	8-18/ 4-8*
16	78.4	0	1.1	0.5		**39.1**	5.8	8-18/ 4-8*
30	62.1	0	1.1	0.5		**31.0**	8.1	8-18
50	25	0	1	0.5		**12.6**	18.4	8-18
100	4.3	6.1	0	0.4		**2.3**	10.3	1.5-5.0
FM Value	2.41	5.94	7.92	6.65	6.33	4.55		

*Lower range acceptable (for any single sieve) if aggregates are slivered, sharp or elongated.

Fig. 2.7 Concrete Mixture 3A.8S with HRWRA and #67 stone

2.5 Concrete Materials and Qualification Testing

Concrete Aggregates Aggregates conformed to ASTM C33, Standard Specification for Concrete Aggregates. The specification defined the requirements for quality of fine and coarse aggregates for use in concrete, including grading, deleterious substances, soundness, and reactivity of aggregate.

Concrete mixtures included 1 1/2-inch nominal maximum size coarse aggregate (size number 467) and 3/4-inch (#67), as defined by ASTM C33, for 12–24-inch-thick concrete members. For members thickness >24 inches, 2-inch (#0.2) coarse aggregate was originally specified. Due to difficulties with local procurement of the larger aggregate and to minimize the number of concrete mixes, a field change request was approved to use one size of the coarse aggregate, # 467 for >24 inches concrete mixtures.

Fig. 2.8 Existing and new powerhouses at Rainbow

Fig. 2.9 Rainbow penstock leading to the powerhouse

Fine and coarse aggregates (source: Ranch Pit) were in compliance with the ASTM C33 requirements for gradation (ASTM C136), material finer than No. 200 sieve (ASTM C117), organic impurities (ASTM C40), lightweight pieces (ASTM C123), clay lumps and friable particles (ASTM C142), and soundness (ASTM C88) tests. The results are presented in Tables 2.17 and 2.18.

Since both the fine and coarse aggregates were tested to be reactive (refer to Chap. 3, ASR Case Study), a low-alkali cement and Class F fly ash were required in concrete mixtures for mitigation of ASR.

Cement The cement, manufactured at the Ash Grove Montana City Plant, conformed to an ASTM C150 Type I/II, with low alkalis and C_3A content. ASTM C1260 expansion tests, made with the cement and the selected aggregates, mitigated the ASR expansion phenomenon associated with the potentially reactive aggregates (refer to Chap. 3 for details). The cement provided a high resistance to the sulfates, present in the local soils. Typical Type II/Type V cement mill test report results are presented in Table 2.19.

Fly Ash Class F fly ash, conforming to ASTM C618, was furnished from Bridger source (Texas). The fly ash had low SO_3 and loss on ignition. Additional testing for the concrete mixes incorporating fly ash was performed on site, with a minimum 20% cement replacement by weight.

Admixtures Air-entraining admixture, Micro Air, manufactured by BASF, conformed to ASTM C260. For the # 67 and # 467 aggregate, the air content required by Table 1.9 (severe exposure) was 6% and 5%, respectively.

Table 2.17 Fine aggregate testing summary (percent passing)

Sieve size	Spec	Ranch Pit
3/8 inch (9.5 mm)	100	100
#4 (4.75 mm)	95–100	99
#8 (2.36 mm)	80–100	80
#16 (1.18 mm)	50–85	57
#30 (600 μm)	25–60	46
#50 (300 μm)	5–30	25
#100 (150 μm)	0–10	4
#200 (75 μm)	0–3	1.6
FM	2.3–3.1	2.89
Soundness loss, %	12	2.3
Absorption, %		0.9
Lightweight pieces, %	0.5 max	0.0
Friable particles, %	3.0 max	0.0
Organic impurities	Color No. 3	Lighter than Color No. 1

NTL Engineering & Geoscience, Inc. Report 91-31, dated August 13, 2007

Table 2.18 Coarse aggregate testing summary

	Specification size No 67 (3/4 in. to No. 4) percent passing sieve		Specification size No. 467 (1 1/2 in. to No. 4) percent passing sieve	
Sieve size	Spec	Ranch Pit	Spec	Ranch Pit
2 in			100	100
1–1/2 in			95–100	100
1 in (25 mm)	100	100		80
3/4 in (19 mm)	90–100	97	35–70	53
1/2 in (12.5 mm)		57		28
3/8 (9.5 mm)	20–55	34	10–30	17
#4 (4.75 mm)	0–10	4	0–5	1.4
#8 (2.36 mm)	0–5	1.0		
#200 (75 μm)				
Specific gravity, SSD				2.61
Absorption, %				1.0
Soundness loss, %	12 max	0.5		
Los Angeles abrasion, % loss (3/4-inch aggregate)	50 max	21		
Material finer than No. 200 sieve, %	1.0	0.3	1.0	0.1
Lightweight pieces, %	0.5 max	0.0	0.5 max	
Friable particles, %	0.5 max	0.0	0.5 max	

NTL Engineering & Geoscience, Inc. Report 91-31, prepared for United Materials, Great Falls, Montana, dated August 13, 2007

Chemical admixtures, conforming to ASTM C494, were furnished by BASF. The following types of chemical admixtures were used:

Polyheed 997, medium range water reducing and retarding admixture, Type A and Type F

Laboratory Trial Mixtures Trial mixture proportions for Class 4000 concrete were selected, in accordance with ACI 211.1. Concrete trial mixes were made, in accordance with ASTM C192, and tested for slump, air content, unit weight, and compressive strength. The aggregate proportions were designed for surface saturated dry (SSD) basis. The trial mixes were targeted to produce 5-inch slump, 5–6 percent air entrainment, and 145 pcf wet unit weight minimum.

Trial batches for 4000 psi concrete with Type II/V cement, sand, #467 coarse aggregate, a mid-range water-reducing admixture, and air-entraining admixture were made with a water-cement ratio, ranging from 0.44 to 0.50. The results of the trial mixes are presented in Table 2.19.

Plastic concrete tests were performed on each trial mix and included slump (ASTM C143), air content (ASTM C231), ambient air and concrete temperatures (ASTM C1064), and unit weight (ASTM C138) measurements (Table 2.20).

Table 2.19 Cement mill test results

Requirements	ASTM C150Type II	Ash GroveType II/V
Chemical requirements		
SiO_2, %	20.0 min.	20.6
Al_2O_3, %	6. 0 max.	3.7
Fe_2O, %	6. 0 max.	3.2
CaO, %		63.7
MgO, %	6. 0 max.	3.2
SO_3, %	3. 0 max.	1.9
LOI, %	3. 0 max.	2.4
Insoluble residue, %	0.75 max.	0.62
Na_2O equivalent, %	0. 60 max.	0.45
C_3S, %		59
C_2S, %		15
C_3A, %	8. 0 max	4.5
C_4AF, %		10
C_4AF+2C_3A, %	25 max.	19
Physical requirements		
Blaine fineness, m^2/kg (air permeability)	280 min	402
Minus 325 mesh		99.60
Vicat set time		
Initial, minutes	45 min.	130
Final, minutes	375 min.	250
Heat of hydration (cal/g) 7 days		92
Air content, %	12 max.	8.1
Autoclave expansion, %	0.80 max	0.04
Compressive strength, psi		
1 day		1970
3 days	1450	3180
7 days	2470	4960
28 days		6640

6 inches by 12 inches cylindrical specimens were cast, in accordance with ASTM C192, for compressive strength tests. The specimens were water cured until tested at 3, 7, and 28 days after casting. Each strength test result was based on an average of three cylinders that were tested at a given age (Figs. 2.10 and 2.11).

Results The selected concrete mixtures for the various structures successfully performed in the field with desired slump and air content with little rejection. For mass concrete placements in the powerhouse, refer to Chap. 5.

Table 2.20 4000 psi laboratory trial mixtures (per cubic yard unless noted)

Designation	Mixture 4000S	Mixture 4000R	Mixture 4000M
Water/cement ratio	0.44	0.44	0.48
Cement type II/V, lb	451	489	388
Fly ash class F, lb	113	122	129
Water, lb	250	266	253
Sand, SSD, lb	1230	1201	1249
#467 stone, SSD, lb	1843	1801	1866
MB Micro Air, oz	9.0	9.8	4.1
MB Polyheed 997, oz	28.2	30.6	25.9
MB Rheobuild 1000, oz	56.4	–	–
Ambient temperature, F	56	50	
Concrete temperature, F	69	68	69
Slump, inches	2/7	5.00	3.00
Air content, %	5.5	5.1	4.8
Unit weight, pcf	143	143.3	143.6
Cylinder compressive strength, psi			
3 days			
7 days	3990	3470	3260
28 days	5060	4970	4310
56 days		5550	

S super-plasticized concrete, *R* regular concrete, *M* mass concrete

Case Study: Cold Weather Construction of a Spillway

The spillway concrete Mixture F4-B was designed for a 28-day compressive strength of 4000 psi. Due to winter concreting, a Type C accelerating admixture, conforming to ASTM C494, was added to concrete. The concrete design mix proportions are shown in Table 2.22.

To reduce the early heat of hydration, 20% of the cement was replaced by fly ash. This reduced cost of the concrete was beneficial from environmental consideration as well.

Mixture FB-4 was used in the chute slabs. The thickness of the spillway chute slab in the upper section was 24 inches and in the lower section was 36 inches. Mixture FB-4M was used in the crest section of the spillway that was 48 inches thick.

Reference 3 provides additional information on concrete mixture design and successful placement under sub-freezing temperatures.

Results The concrete mixtures were successfully placed and placed in the cold weather without any quality issues.

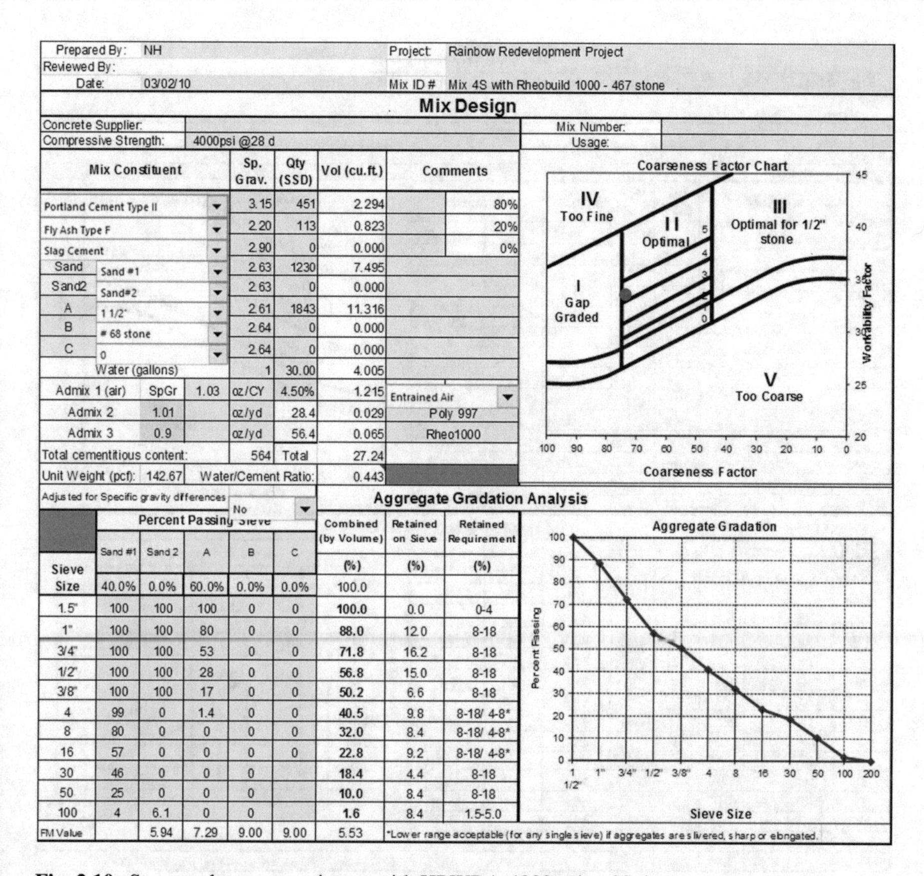

Prepared By:	NH	Project	Rainbow Redevelopment Project
Reviewed By:			
Date:	03/02/10	Mix ID #	Mix 4S with Rheobuild 1000 - 467 stone

Mix Design

Concrete Supplier:		Mix Number:	
Compressive Strength:	4000psi @28 d	Usage:	

Mix Constituent			Sp. Grav.	Qty (SSD)	Vol (cu.ft.)	Comments	
Portland Cement Type II			3.15	451	2.294		80%
Fly Ash Type F			2.20	113	0.823		20%
Slag Cement			2.90	0	0.000		0%
Sand	Sand #1		2.63	1230	7.495		
Sand2	Sand#2		2.63	0	0.000		
A	1 1/2"		2.61	1843	11.316		
B	# 68 stone		2.64	0	0.000		
C	0		2.64	0	0.000		
Water (gallons)			1	30.00	4.005		
Admix 1 (air)	SpGr	1.03	oz/CY	4.50%	1.215	Entrained Air	
Admix 2	1.01		oz/yd	28.4	0.029	Poly 997	
Admix 3	0.9		oz/yd	56.4	0.065	Rheo1000	
Total cementitious content:			564	Total	27.24		
Unit Weight (pcf):	142.67	Water/Cement Ratio:			0.443		

Adjusted for Specific gravity differences: No

Aggregate Gradation Analysis

	Percent Passing Sieve					Combined (by Volume)	Retained on Sieve	Retained Requirement
Sieve Size	Sand #1	Sand 2	A	B	C	(%)	(%)	(%)
	40.0%	0.0%	60.0%	0.0%	0.0%	100.0		
1.5"	100	100	100	0	0	100.0	0.0	0-4
1"	100	100	80	0	0	88.0	12.0	8-18
3/4"	100	100	53	0	0	71.8	16.2	8-18
1/2"	100	100	28	0	0	56.8	15.0	8-18
3/8"	100	100	17	0	0	50.2	6.6	8-18
4	99	0	1.4	0	0	40.5	9.8	8-18/ 4-8*
8	80	0	0	0	0	32.0	8.4	8-18/ 4-8*
16	57	0	0	0	0	22.8	9.2	8-18/ 4-8*
30	46	0	0	0		18.4	4.4	8-18
50	25	0	0	0		10.0	8.4	8-18
100	4	6.1	0	0		1.6	8.4	1.5-5.0
FM Value		5.94	7.29	9.00	9.00	5.53		

*Lower range acceptable (for any single sieve) if aggregates are slivered, sharp or elongated.

Fig. 2.10 Structural concrete mixture with HRWRA-4000 psi at 28 days

2.6 Conclusions and Recommendations

The following conclusions are based on the results and analyses presented in this chapter:

1. Perform testing of concrete trial mixtures in accordance with ACI 211.1 with proper selection of concrete materials and admixtures, etc. The w/cm and design compressive strength should be in accordance with ACI 318 for the respective class and exposure condition.
2. Allow adequate lead time for performing the trial mixtures, prior to construction.
3. Verify the results of the selected trial mixtures in the field, using the proposed batching facility, including slump, air content, and compressive strength.
4. Establish the dosage of air-entraining and chemical admixtures, based on slump, air content, and setting time for the anticipated range of the site conditions.

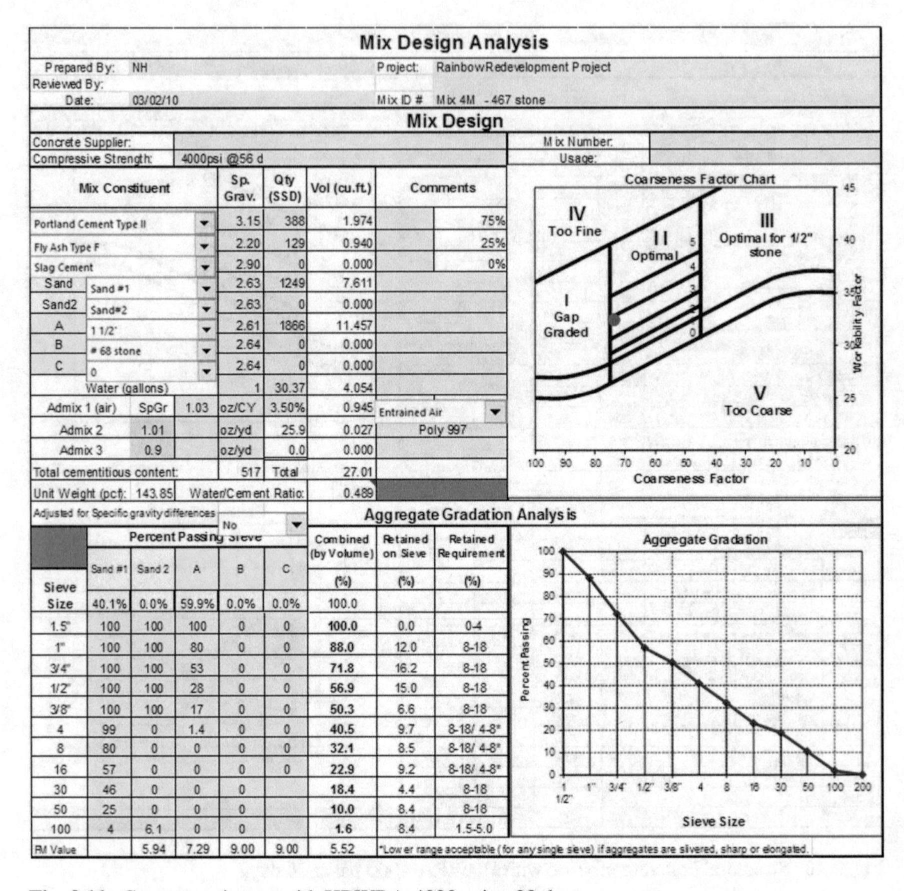

Mix Design Analysis

Prepared By: NH
Reviewed By:
Date: 03/02/10
Project: RainbowRedevelopment Project
Mix ID # Mix 4M - 467 stone

Mix Design

Concrete Supplier:
Compressive Strength: 4000psi @56 d
Mix Number:
Usage:

Mix Constituent		Sp. Grav.	Qty (SSD)	Vol (cu.ft.)	Comments
Portland Cement Type II		3.15	388	1.974	75%
Fly Ash Type F		2.20	129	0.940	25%
Slag Cement		2.90	0	0.000	0%
Sand	Sand #1	2.63	1249	7.611	
Sand2	Sand#2	2.63	0	0.000	
A	1 1/2"	2.61	1866	11.457	
B	# 68 stone	2.64	0	0.000	
C	0	2.64	0	0.000	
Water (gallons)		1	30.37	4.054	

Admix	SpGr		Unit	Qty	Vol (cu.ft.)	Comments
Admix 1 (air)	SpGr	1.03	oz/CY	3.50%	0.945	Entrained Air
Admix 2	1.01		oz/yd	25.9	0.027	Poly 997
Admix 3	0.9		oz/yd	0.0	0.000	

Total cementitious content: 517 Total 27.01
Unit Weight (pcf): 143.85 Water/Cement Ratio: 0.489
Adjusted for Specific gravity differences: No

Aggregate Gradation Analysis

Sieve Size	Sand #1	Sand 2	A	B	C	Combined (by Volume) (%)	Retained on Sieve (%)	Retained Requirement (%)
	40.1%	0.0%	59.9%	0.0%	0.0%	100.0		
1.5"	100	100	100	0	0	100.0	0.0	0-4
1"	100	100	80	0	0	88.0	12.0	8-18
3/4"	100	100	53	0	0	71.8	16.2	8-18
1/2"	100	100	28	0	0	56.9	15.0	8-18
3/8"	100	100	17	0	0	50.3	6.6	8-18
4	99	0	1.4	0	0	40.5	9.7	8-18/ 4-8*
8	80	0	0	0	0	32.1	8.5	8-18/ 4-8*
16	57	0	0	0	0	22.9	9.2	8-18/ 4-8*
30	46	0	0	0		18.4	4.4	8-18
50	25	0	0	0		10.0	8.4	8-18
100	4	6.1	0	0		1.6	8.4	1.5-5.0
FM Value		5.94	7.29	9.00	9.00	5.52		

*Lower range acceptable (for any single sieve) if aggregates are silvered, sharp or elongated.

Fig. 2.11 Concrete mixture with HRWRA-4000 psi at 28 days

Table 2.21 Rainbow 4000 psi compressive strength concrete quantities and costs

Structure	Quantities (cy)	Unit cost ($/cy)
Powerhouse mass concrete	13170	415
Powerhouse walls, floors, etc.	5540	450–600
Power canal intake	3530	608
Power canal walls and SOG	15905	516
Forebay	1765	648
Spillway ogee mass	3315	333
Spillway walls	1335	550
Penstock intake	3530	461

Table 2.22 Concrete design mixture proportions (lbs/yd^3 unless noted)

Material	Source	Mixture FB-4 with 3/4-inch aggregate	Mixture FB-4 with 1–1/2-inch aggregate
Cement	Type II low-alkali	470	451
Fly Ash	Type F	118	113
Coarse aggregate	Engman Pit ¾″	1800	1200
Coarse aggregate	Engman Pit 1–1/2″	–	600
Fine aggregate	Engman Pit Sand	1325	1425
AEA	BASF-AE 90	3.5 oz/cwt	3.5 oz/cwt
WRDA	BASF-Polyheed 1020	14.7 oz./cy	16.9 oz./cy
Accelerating admixture	BASF-NC 534	20 oz./cwt[a]	20 oz./cwt[a]
Water	Well	259	248
Water/cementitious material ratio		0.44	0.44

[a]NC 534 admixture was added during cold weather to accelerate the setting time and to increase the early strength of concrete

5. For high-performance concrete, incorporate supplementary cementitious materials and specialty admixtures, such as SRA, to reduce drying shrinkage for optimum results.
6. For concrete exposed to freezing and thawing cycles, the air content requirement for durability of individual concrete design mixes should be defined by the size of aggregate and weathering region where the project is located and should follow the ACI guidelines. Perform ASTM C666 tests on determine the resistance of concrete mixtures to freeze-thaw cycles.
7. For cold weather concreting, incorporate an accelerating admixture, Type C, ASTM C494, for controlling concrete workability and early strength development.

References

1. Hasan N (2005) Performance of concrete at San Roque dam auxiliary structures. In: Proceedings of Third International Conference on Construction Materials and Mindness Symposium, Vancouver, Canada
2. Hasan N (1991) Superplasticized concrete for a prefabricated powerplant structure. In: ASCE Waterpower 1991 Proceedings of the International Conference, Denver, Colorado, Volume 3
3. Hasan N (2016) Sustainability and durability of concrete placed in cold weather. In: 2016 International Concrete Sustainability Conference, Washington, DC

Chapter 3
Alkali-Silica Reactivity Mitigation

3.1 General

The risk of damaging expansion caused by alkali-aggregate reaction (AAR) in concrete is well known. The risk of damage comes from alkali-silica reactive aggregates or alkali carbonate reactive aggregates. In both cases, it causes premature damage in concrete. The former is of higher concern since aggregates containing various forms of silica materials are very common, whereas the latter occurs rarely because of the unsuitability of carbonates for use in concrete. Nonetheless, concrete deterioration caused by each type of alkali-aggregate reaction is similar.

The alkali-silica reaction (ASR) is a well-known phenomenon, causing premature distress in concrete structures and pavements, shortening their service life. It was first recognized as a problem in North America as reported by Stanton [1] in 1940, in Denmark in the early 1950s, in West Germany in the early 1960s, in the UK in the mid-1970s, and in Japan in the early 1980s. The first case of alkali-carbonate reaction (ACR) was observed in the late 1950s by Swenson [2].

Figure 3.1 shows that the phenomenon of ASR is widespread within the USA and may reduce service life of concrete in the absence of mitigation measures. Confirmed cases of ACR have been limited to a few locations in the USA: Virginia, Indiana, Iowa, and Illinois.

It should be known that no structure has ever collapsed due to alkali-aggregate reactions, but there are cases in which structural concrete members were demolished due to the effect of alkali-aggregate reactions.

Most of the structures severely cracked by ASR are exposed to the weather or are in contact with damp soil. This is because, for a significant amount of expansion to occur, sufficient presence of moisture is essential. Apart from moisture, high content of alkali in the concrete is also essential (Fig. 3.2).

Reference [3] provides a world review of alkali-aggregate reactivity in concrete. Reference deals with the most recent findings concerning the mechanisms involved in the reaction and methods concerning its diagnosis, testing, and evaluation, together with an appraisal of current methods used in its avoidance and in the

N. Hasan, *Durability and Sustainability of Concrete*,
https://doi.org/10.1007/978-3-030-51573-7_3

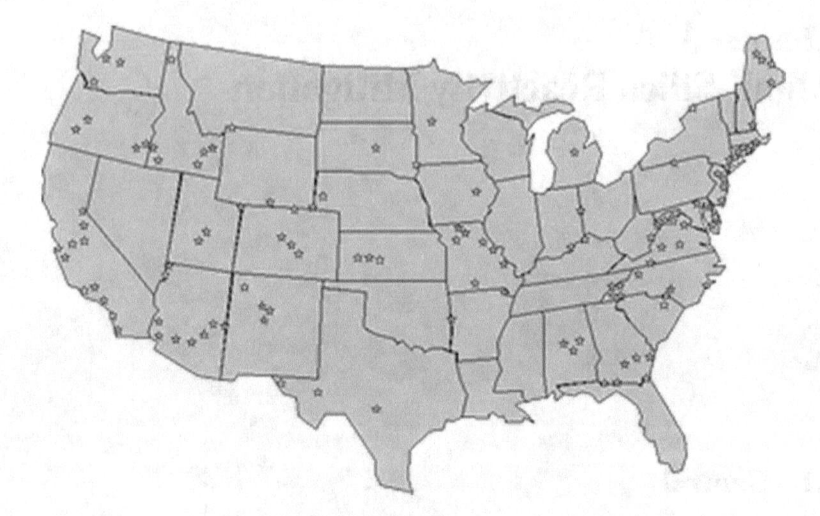

Fig. 3.1 Location of structures in the USA known to be affected by the alkali-aggregate reaction in concrete, sources of sand and gravel subject to alkali-silica reaction and alkali-carbonate reaction (Courtesy of FHWA Report 2003, Ref. [8])

Fig. 3.2 Typical ASR distress in concrete with reactive aggregates exposed to moisture (Ref. [8])

remediation of affected concrete structures. It includes chapters covering each region of the world, written by experts with specialist knowledge of AAR, and includes an authoritative appraisal of the problem and its solution as it affects concrete structures in the region.

Reference [4] is a comprehensive report on the alkali-silica reaction on nuclear plant structures, published by the Electric Power Research Institute (EPRI) in 2017. This includes the following abstract:

"Nuclear plants with concrete structures affected by alkali-silica reaction (ASR) need to develop aging management programs in order to manage the deterioration and to ensure their integrity for extended operations. This report provides a literature review of research performed on ASR affected structures and highlights the changes in mechanical properties and structural behavior when ASR is present. ASR can affect concrete structures in different ways depending on the orientation of reinforcing steel in the structure and the intended function of the structure itself."

The ASR is of higher concern since aggregates containing various forms of silica materials are very common, whereas the latter occurs rarely because of the unsuitability of carbonates for use in concrete.

Most of the structures severely cracked by ASR are exposed to the weather or are in contact with damp soil. This is because, for a significant amount of expansion to occur, sufficient presence of moisture is essential. Apart from moisture, high content of alkali in the concrete is also essential.

3.2 Types of Alkali-Aggregate Reaction

Alkali-Silica Reaction (ASR) Alkali-silica reaction (ASR) is a chemical reaction between the alkali hydroxide in Portland cement and certain siliceous rocks and minerals present in aggregates, such as opal, chert, chalcedony, trimite, cristobalite, strained quartz, etc. The resulting alkali-silica gel then absorbs water and expands, which results in cracking of the aggregates, the cement paste, and ultimately the concrete matrix. Random map cracking and closed joints and attendant spalling concrete are indicators of alkali-silica reactions.

The ASR needs several components to take place: Adequate amounts of alkali (provided by cement), reactive silica aggregate, and moisture must be present for significant damage to concrete.

Alkali-silica reaction (ASR) results in random map cracking and attendant spalling concrete. Cracking initiates in areas with a frequent supply of moisture, such as in the proximity of waterline in piers and near the ground behind retaining walls. Petrographic examination can identify alkali-silica reactions. It can be controlled using proper portions of supplementary cementitious materials like silica fume, fly ash, and ground-granulated blast-furnace slag or admixtures such as lithium compounds.

Alkali-Carbonate Reaction (ACR) ACR occurs between alkali hydroxides and certain argillaceous dolomite limestones. The reaction results in the expansion and cracking of concrete. It is observed with certain dolomitic rocks. The reaction is similar to the ASR except no visible gel is formed. Compared to alkali-silica reactions, ACR is fairly rare because aggregates susceptible to this phenomenon are less common. The use of supplementary cementing materials does not prevent deleterious expansion due to ACR. Rocks potentially susceptible are dolomite limestones containing 40–60% of the total carbonate fraction of the rock, in which there is a 10–20% clay fraction and in which a texture of small dolomite crystals is scattered throughout a matrix of extremely fine-grained calcite and clay.

3.3 Sources of Alkalis in Concrete

1. *Cement*: All ingredients of concrete may contribute to the total alkali content of the concrete; the major source of alkali is from cement, oxides of sodium and potassium oxides ($Na_2O + 0.658K_2O$).
2. *Aggregate*: Aggregate containing feldspars, some micas, glassy rock, and glass may release alkali in concrete. Sea-dredged sand, if not properly washed, may contain sodium chloride which can contribute significant alkali to concrete.
3. *Admixtures*: Admixture in the context of ASR in concrete means chemical agents are added to concrete at the mixing stage. These include accelerators, water reducers (plasticizers), retarders, superplasticizers, air-entraining agents, etc. Some of the chemicals contain sodium and potassium compounds which may contribute to the alkali content of concrete.
4. *Water*: Water may contain alkalis in solution.
5. *External Sources of Alkalis*: In cold weather conditions, the use of deicing salt containing sodium compounds may increase the alkali content on the surface layer of concrete. Soils containing alkali may also increase alkali content on the surface of concrete.

3.4 Effects of Alkali-Aggregate Reaction

1. Loss of strength, stiffness, impermeability.
2. Affects concrete durability and appearance.
3. Premature failure of concrete structures.
4. Consequently, life of concrete structure is declined.
5. Maintenance cost is increased.

3.5 Alkali-Aggregate Reactivity Tests

ASTM has several test methods to identify potentially reactive aggregates. ASTM C295 is standard guide for petrographic examination of aggregates. This quick test would alert the engineer to the presence of potentially reactive rock and mineral type. Table 3.1 lists the ASTM test methods to evaluate alkali-silica reactivity (ASR) and alkali-carbonate reactivity (ACR) and their test duration and reliability.

Table 3.1 ASTM test methods for evaluating alkali-aggregate reactivity

Test method	Test for	Type/duration	Reliability	Remarks
ASTM C289	ASR	Chemical/24 hours	Poor	Withdrawn in 2016
ASTM C295	ASR	Petrographic analyses	Good	Rocks/minerals potentially reactive, preliminary screening
ASTM C227	ASR	Expansion/3, 6, and 12 months	Good	Withdrawn in 2018
ASTM C441	ASR	Expansion/14 days/1 year	Variable	Efficacy of SCM
ASTM C1260	ASR	14 days or 28 days	Good	Run with high-alkali cement
ASTM C1293	ASR	Expansion/1 year/2 years	Most reliable	Long lead time
ASTM C1567	ASR	Expansion/14 days or 28 days	Good	Run with job cement and aggregates
ASTM C586	ACR	Expansion	Good	Screening test for rocks/minerals
ASTM C856	ASR	Petrographic analyses	Good	For hardened concrete
ASTM C1105	ACR	Expansion/3 months, 6 months, 12 months	Good	Long lead time

ASTM C289 was a quick test where the aggregate particles are immersed in a NaOH solution in sealed container for 24 hours at 80 °C. The resulting suspensions are filtered and analyzed. A curve is then drawn between the reduction in alkalinity and dissolved silica on a log scale. Innocuous aggregates are considered to be those that produce little or no reduction in alkalinity. Deleterious or potentially deleterious materials are identified as those with loss reduction in alkalinity, as shown in Fig. 3.3. This test, due to its limitations and poor reliability, was withdrawn by ASTM in 2016.

ASTM C1260 Test ASTM C1260 is an accelerated mortar bar test. In this test method, mortar prisms, measuring 1 × 1 × 11.25 inches (25 × 25 × 285 mm), are cast and stripped after 1 day, and initial length of the samples is measured. The mortar bars are then stored in 1 molar NaOH solution at 176 °F (80 °C) during the test period. The average14-day expansion below 0.1% expansion indicates innocuous behavior, whereas an expansion more than 0.20% indicates potentially deleterious behavior.

ASTM C1293 Test ASTM C1293 is a mortar bar test, but most reliable, for detecting deleterious ASR behavior of an aggregate using a combination of pozzolan or slag aggregate in 1 year. In this test method, concrete prisms are cast using a high-alkali cement and stripped after 1 day, and initial length of the samples is measured. Specimens are then stored during the test period. Additional NaOH is added to the mixing water to provide alkali loading 8.85 lb/yd^3 (5.25 kg/m^3). The average 1-year expansion below 0.04% expansion indicates innocuous behavior. For evaluation of SCM and chemical admixtures, run this test for 2 years to verify expansion.

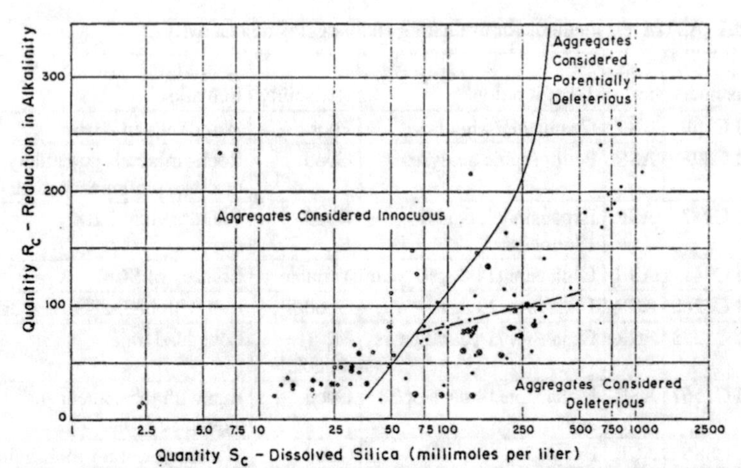

FIG. 2 Illustration of Division Between Innocuous and Deleterious Aggregates on Basis of Reduction in Alkalinity Test

Fig. 3.3 ASTM C289 tests for ASR based on dissolved silica (test withdrawn)

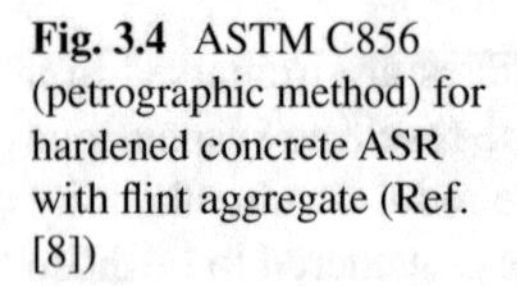

Fig. 3.4 ASTM C856 (petrographic method) for hardened concrete ASR with flint aggregate (Ref. [8])

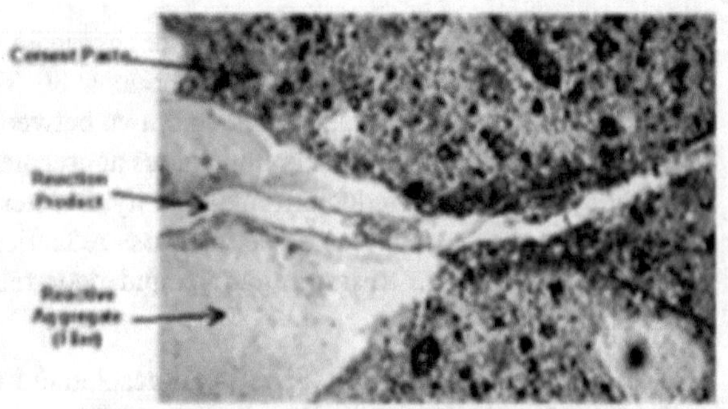

ASTM C1105 This test is similar to ASTM C1293 except for storage and alkali loading. Specific concrete mixture specimens are stored 73 °F (23 °C) for up to a year. There is no addition of NaOH. Potentially deleterious expansion limits are 0.015% at 3 months, 0.025% at 6 months, and 0.03% at 12 months.

ASTM C1567 Test ASTM C1567 [3] is an accelerated mortar bar test for detecting deleterious ASR behavior of an aggregate using a combination of hydraulic cement, supplementary cementitious materials, and aggregate within 16 days. The method is similar to ASTM C1260, except for the combinations of the proposed cementitious materials for the project. The average 14-day expansion below 0.1% expansion indicates innocuous behavior, whereas an expansion more than 0.10% indicates potentially deleterious expansion and further testing by ASTM C1293 (Fig. 3.4).

3.6 Mitigation of ASR

General There are a number of mitigation measures that can be taken to minimize the potential of damage due to ASR, which may include:

1. Using non-reactive aggregates
2. Limiting the alkali content of the concrete
3. Incorporating supplementary cementitious materials (SCMs) such as Class F or Class C fly ash, GGBF slag, or silica fume
4. Incorporating chemical admixtures, such as lithium compounds

Using non-reactive aggregates in concrete should be the first preference in mitigation of ASR or ACR. This can be based on the proven service record of concrete incorporating the aggregate from the known quarry or river-run materials. If a service record is not available or not representative of the environment to which the proposed concrete will be exposed, accelerated tests should be made with the proposed aggregate to verify innocuous behavior.

A good option is to specify a low-alkali cement (less than 0.60% equivalent Na_2O) to avoid ASR. This should be verified by ASTM C1567 test using the proposed cement-aggregate combination.

Prior to 1970 using low-alkali cement, those with 0.6% or less sodium oxide equivalent, was believed to adequately mitigate ASR in concrete. In the mid-1970s, 7-year-old concrete pavement in Albuquerque showed significant deterioration. The PCA diagnosed the distress as a result of ASR. The solution recommended and implemented was to introduce fly ash as a partial cement replacement to further reduce alkali content in concrete mixes.

In the 1960s, USBR Concrete Manual [5] considered that a low-alkali (0.6% equivalent sodium oxide) cement alone was effective in mitigating ASR in concrete. This criterion was widely adopted by the industry and other agencies, including ACI, FHA, and ASTM. The ASR reactivity potential was evaluated by the conventional test methods, such as ASTM C227 (mortar bar test), which was timc-consuming. More recent studies by Stark of USBR [6] and others have shown that 0.6% alkali may not prevent ASR damage to concrete in every case. Certain slowly reacting aggregates, such as quartzite, may slowly react to lower-alkali cements under moist environmental conditions. Also, alkalis from other sources in concrete mixture may elevate the alkali content to significant levels even if low-alkali cement is used. In addition to cement, alkali content from the aggregates, admixtures, and external sources, such as water and deicing salts during service conditions, must be considered to predict the long-term field performance of ASR in concrete. This has led the industry and testing agencies to develop better methods to mitigate the effects of ASR in concrete.

In 1997, the New Mexico State Department of Transportation (NMDOT) initiated a research project to conduct experiments to ascertain the level and type of additives, such as fly ash and lithium nitrates, required to reduce expansion to acceptable levels. The work was conducted at the ATR Institute, University of New Mexico, and the Research Bureau, NMDOT.

Reference [7] has published these results of research experiments aimed at the mitigation of alkali-silica reactivity in New Mexico. Most of the reactive aggregates in the State of New Mexico are found in the western part of the State. The aggregates in the southeastern part of the State are generally of low reactivity or non-reactive.

The study concluded that a minimum of 25% fly ash by weight of total cementitious material should be added in concrete to mitigate the ASR distress. An acceptable alternative is a combination of lithium nitrate at the manufacturer's recommended dosage plus fly ash at a minimum of 15% by weight of total cementitious material.

Rangaraju and Sompura [9] in their 2005 paper have presented the results of a laboratory study conducted to determine the influence of cement composition and fly ash on the ASR expansions observed in standard and modified ASTM C1260 tests. The results indicated that the use of low-alkali cement in combination with Class F and Class C fly ashes yielded larger reductions in percent expansion than the cement alone; also, Class F fly ash is more effective than Class C fly ash in mitigating the expansion. If Class C fly ash is used to mitigate ASR expansion, [10] recommends a minimum of 40% fly ash replacement of cement to reduce the 14-day expansion below 0.10%.

A typical chemical analysis of Headwaters Resources fly ash is presented in Fig. 3.12.

Use of Class F fly ash, as supplement to the low-alkali cement, to prevent expansive alkali-silica reaction is recommended. An optional requirement for Class F fly ash is to limit its alkali content to 1.5% maximum. Class C fly ash, which has usually higher lime (CaO) content than Class F, is considered less effective to prevent ASR unless a higher replacement of cement is used. ASTM C441 is used to evaluate fly ash with a 75% reduction in expansion at 14 days.

If the ASR expansion exceeds 0.1% and confirms that the low-alkali cement alone will not prevent deleterious expansion, using a SCM is recommended. The expansion characteristics of the proposed aggregate, cement, and fly ash combinations should be further verified by ASTM C1567 test prior to use (Fig. 3.5).

Other SCMs to prevent expansive ASR include slag cement (GGBFS) and silica fume; ASTM C441 is used to evaluate slag cement. Slag cement should be tested in accordance with ASTM C989 for a reduction in expansion to at least 75%. Figure 3.6 shows the effectiveness of slag cement to mitigate ASR expansion, when added at 30–50% replacement of cement.

Silica fume, if used to prevent ASR expansion, should replace the cement by up to 10%.

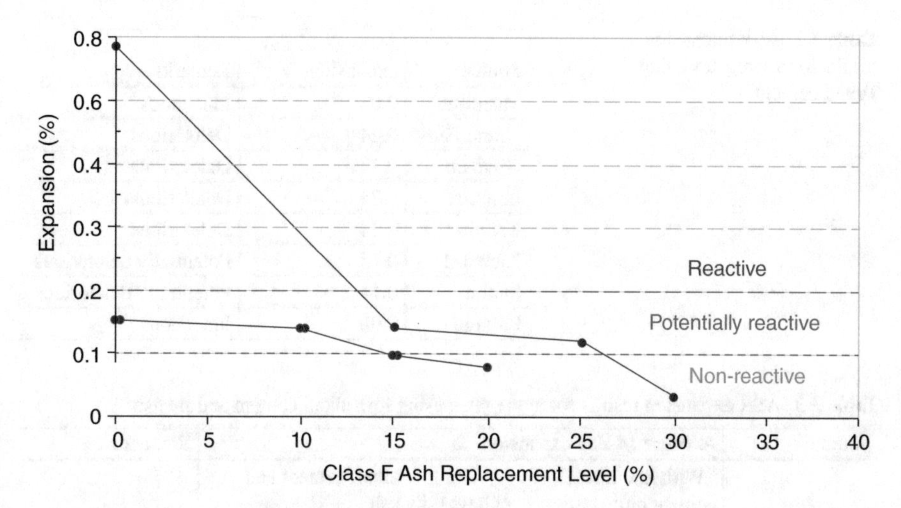

Fig. 3.5 ASTM C1260 expansion with Class F fly ash replacement levels (Courtesy of BASF)

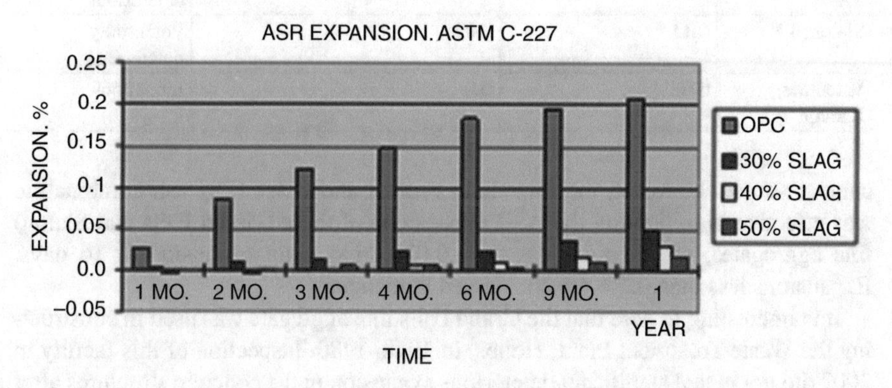

Fig. 3.6 ASTM C227 ASR expansion test showing effect of slag cement on ASR (Courtessy of BASF)

Case Study: National Enrichment Facility (NEF), Lea County, New Mexico (2007) [11]

At the National Enrichment Facility, New Mexico, several local fine and coarse aggregates in combination with standard Type I cement were tested in accordance with ASTM C1260 test method [9]. Most aggregates exhibited deleterious ASR expansion (>0.1% at 14 days). The results are shown in Table 3.2.

The use of the project cement with low alkalis (0.5%) reduced the ASR expansion to 0.19% at 14 days. However, the use of ASTM C618 Class F fly ash, with 20% replacement of the cement in concrete, resulted in mitigating the ASR expansion to less than 0.1%. The results are presented in Table 3.3.

The available fine aggregate sources from Grand Falls and Slaton and Valentine in New Mexico were found to be potentially deleterious, even with a low-alkali

Table 3.2 ASR expansion results for aggregates using Type I cement

Source	Average 14-day expansion, %	Remarks
Valentine	0.03	Innocuous
Grand Falls	0.24	Deleterious
Grisham	0.27	Deleterious
Imperial	0.24	Deleterious
Kiewit-1	0.24	Deleterious
Kiewit-2	0.15	Potentially deleterious
Slaton	0.14	Potentially deleterious
Lafarge	0.08	Innocuous

Table 3.3 ASR expansion results for aggregates using low-alkali cement and fly ash

Source	Average 14-day expansion, %		Remarks
	With low-alkali cement only	With low-alkali cement and Class F fly ash	
Grand Falls, TX	0.19	0.01	Potentially deleterious
Slaton, TX	0.11	0.02	Potentially deleterious
Valentine, Lafarge	0.02	0.0	Innocuous

cement source. However, the low-alkali cement and Class F fly ash combination was effective in mitigating the ASR expansions of these (Grand Falls and Slaton) fine aggregates, resulting in less than 0.02% maximum expansion at 16 days. Expansions less than 0.1% are considered innocuous.

It is interesting to note that the Grand Falls fine aggregate was used in constructing the Waste Treatment Plant, Hobbs, in 1978–1980. Inspection of this facility in 2007 did not reveal significant deleterious expansion in the concrete structures after 27 years.

Results As a follow-up to the ASR test results, it was decided to include Class F fly ash in all concrete mixes for the NEF Project to mitigate the ASR distress in concrete. A combination of a low-alkali (0.45%) Cemex cement with a 20% minimum cement replacement (by weight) by fly ash was used in all concrete.

Case Study: Hebgen Project, Butte, MT (2010)

Since local fine aggregate and coarse aggregate were potentially reactive, tests were performed in accordance with ASTM C1260 and ASTM C1567 to verify the 16-day expansion. All tests were performed with low-alkali cement (0.5% alkalis) and up to 25% fly ash replacement of cement as shown in Figs. 3.7, 3.8, 3.9, 3.10. and 3.11.

Project # 0 ASTM C-1260 Potential Alkali Reactivity

Project: ASR Testing

Identification: 100% Fine

Date Cast: March 24, 2010

Coarse Agg Source:
Percent Coarse Agg.

Fine Agg Source: BASF High Caliber Concrete
Percent Fine Agg. 100%

Cement Source & Type BASF
Test Type:

Fly Ash Source:
Curing Temp. = 176+_3.6F

Fly Ash Percentage: Replacement
Alkali (Na2O) = 0.50%

Water/Cement Ratio = 0.47
Lab#: 0

Age in Days	Age in NaOH	Percent Change	Specification	Date Measured
2	0	0.00		26-Mar
7	5	0.17		31-Mar
12	10	0.30		5-Apr
16	14	0.35	0.1	9-Apr

Fig. 3.7 ASTM C1260 ASR test with 100% fine aggregate

It is evident that both the low-alkali cement and Class F fly ash combinations were required in mitigating the ASR expansions of the local aggregates, to 0.05% maximum expansion at 14 days. Expansions less than 0.1% are considered innocuous.

Results As a follow-up to the ASR test results, it was decided to include Class F fly ash with a 25% minimum cement replacement (by weight) for the Hebgen concrete mixtures.

Chemical analysis of Headwaters Class C and Class F fly ash, used in the above study, is presented in Fig. 3.12.

Project # 0 ASTM C-1260 Potential Alkali Reactivity

Project: ASR Testing

Identification: 100% Coarse

Date Cast: March 24, 2010

Coarse Agg Source: BASF High Caliber Concrete
Fine Agg Source:
Cement Source & Type BASF
Fly Ash Source:
Fly Ash Percentage: Replacement
Water/Cement Ratio = 0.47

Percent Coarse Agg. 100%
Percent Fine Agg.
Test Type:
Curing Temp. = 176+_3.6F
Alkali (Na2O) = 0.50%
Lab#: 0

Age in Days	Age in NaOH	Percent Change	Specification	Date Measured
2	0	0.00		26-Mar
7	5	0.12		31-Mar
12	10	0.27		5-Apr
16	14	0.33	0.1	9-Apr

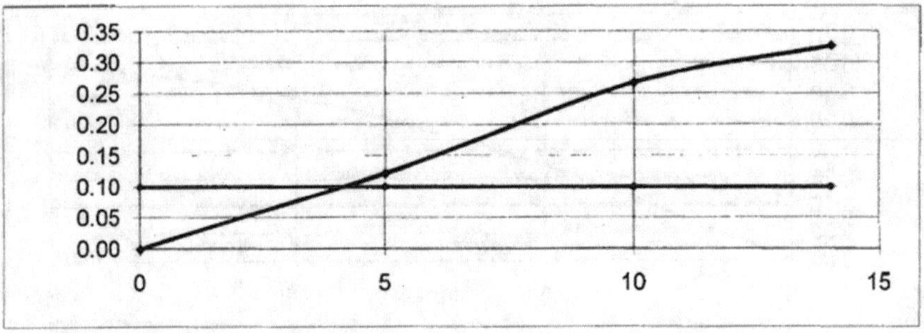

Fig. 3.8 ASTM C1260 ASR test with 100% coarse aggregate

Case Study: Rainbow Redevelopment Project, Great Falls, MT (2009–2011)

The concrete structures at the project site in Great Falls, Montana, are exposed to cold and wet environmental conditions. The local sand and coarse aggregate were initially tested in accordance with ASTM C1260 test method and exhibited ASR expansion of 0.13% and 0.21% at 14 days, respectively. Use of ASTM C618 Class C fly ash, with 20% replacement of cement, tested in accordance with ASTM C1567 test method, exhibited ASR expansion of 0.16–0.24% at 16 days. Subsequently, ASTM C618 Class F fly ash, with 20% replacement of the cement in concrete, was used to mitigate the ASR expansion to less than 0.05%. The results are shown in Tables 3.4 and 3.5.

Additional ASTM C1260 testing of Ranch Pit fine and coarse aggregate with the low-alkali project cement (Ash Grove) and Class F fly ash (Headwaters) was performed at the laboratory. The results are presented in Table 3.5.

Project # 0 ASTM C-1567 Potential Alkali Reactivity

Project: Lab Services

Identification: 60/40 Blend 20% Ash

		Date Cast:	August 23, 2010
Coarse Agg Source:	High Caliber	Percent Coarse Agg.	60%
Fine Agg Source:	High Caliber	Percent Fine Agg.	40%
Cement Source & Type	Holcim I / II	Test Type:	
Fly Ash Source:	Cold Creek	Curing Temp. =	176+_ 3.6F
Fly Ash Percentage:	20% Replacement	Alkali (Na2O) =	0.50%
Water/Cement Ratio =	0.47	Lab#:	0

Age in Days	Age in NaOH	Percent Change	Specification	Date Measured
2	0	0.00		25-Aug
7	5	0.02		30-Aug
12	10	0.04		4-Sep
16	14	0.05	0.1	8-Sep

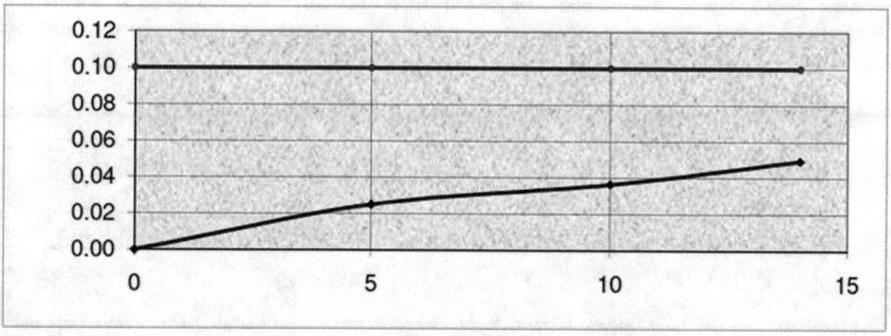

Fig. 3.9 ASTM C1567 ASR test with 60:40 aggregate blend and 20% fly ash replacement

It is evident from Table 3.4 that the low-alkali cement and Class F fly ash combination is really effective in mitigating the ASR expansions of the Ranch Pit aggregates, with 0.05% maximum expansion. Expansions less than 0.1% are considered innocuous.

Cement The cement, manufactured at the Ash Grove Montana City Plant, conformed to an ASTM C150 Type II/V, with low alkalis and C_3A content. ASTM C1260 expansion tests, made with Cemex cement and the selected aggregates, mitigated the ASR expansion phenomenon associated with the potentially reactive aggregates (Table 3.4). The cement provides a high resistance to the sulfates, present in the local soils. Typical Type II/Type V cement mill test report results are presented in Table 3.6.

Fly Ash Class F fly ash, conforming to ASTM C618, was furnished from Bridger source (Texas). The fly ash has low SO_3 and loss on ignition. A certified test report mill for the Class F fly ash is shown in Fig. 3.12: Additional testing for the concrete mixes incorporating fly ash was performed on site, with a minimum 20% cement replacement by weight.

Project # 0 ASTM C-1567 Potential Alkali Reactivity

Project: Lab Services

Identification: 60/40 Blend 25% Ash

		Date Cast:	August 23, 2010
Coarse Agg Source:	High Caliber	Percent Coarse Agg.	60%
Fine Agg Source:	High Caliber	Percent Fine Agg.	40%
Cement Source & Type	Holcim I / II	Test Type:	
Fly Ash Source:	Cold Creek	Curing Temp. =	176+_ 3.6F
Fly Ash Percentage:	25% Replacement	Alkali (Na2O) =	0.50%
Water/Cement Ratio =	0.47	Lab#:	0

Age in Days	Age in NaOH	Percent Change	Specification	Date Measured
2	0	0.00		25-Aug
7	5	0.03		30-Aug
12	10	0.04		4-Sep
16	14	0.05	0.1	8-Sep

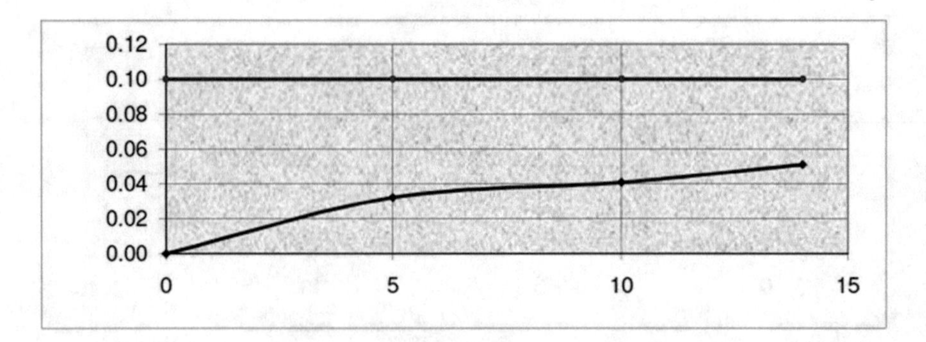

Figure 3.10 ASTM C1567 test run with 60/40 aggregate blend and 25% fly ash replacement

Results As a follow-up to the ASR test results, it was decided to include Class F fly ash in all concrete mixes for the project to mitigate the ASR distress for project. Considering the low alkalis (0.45%) of cement, a 20% minimum cement replacement (by weight) by fly ash was recommended.

Case Study: San Roque Multipurpose Project, Philippines (1998–2002)

Table 3.7 shows the ASR results for Mixes 1–6. Mixes 5 and 6, made with the project aggregates, project cement, and 20% fly ash replacement, exhibited the least expansion, at 14-day and 28-day ages.

Results Based on the test results, all concrete mixtures for the San Roque Project were designed with a low-alkali cement and a Class F fly ash to mitigate ASR.

Project # ASTM C-1567 Potential Alkali Reactivity

Project: BASF High Caliber Concrete

Identification: 55/45 Blend w/25% Coal Creek

		Date Cast:	May 10, 2010
Coarse Agg Source:	BASF High Caliber Co	Percent Coarse Agg.	55%
Fine Agg Source:	BASF High Caliber Co	Percent Fine Agg.	45%
Cement Source & Type	High Caliber Concrete	Test Type:	
Fly Ash Source:	Coal Creek	Curing Temp. =	176+_ 3.6F
Fly Ash Percentage:	25% Replacement	Alkali (Na2O) =	0.50%
Water/Cement Ratio =	0.47	Lab#:	0

Age in Days	Age in NaOH	Percent Change	Specification	Date Measured
2	0	0.00		12-May
7	5	0.01		17-May
12	10	0.01		22-May
16	14		0.1	26-May

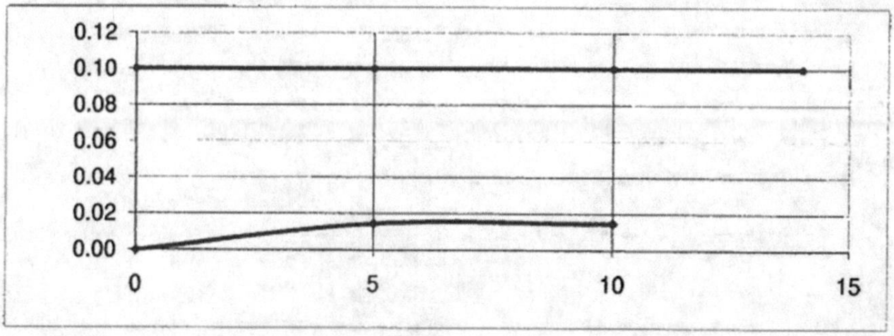

Fig. 3.11 ASTM C1567 tests with 100% with 55/45 aggregate combination and 25% fly ash replacement

3.7 Chemical Admixtures

Chemical admixtures, such as lithium salts, are known to inhibit deleterious expansion due to reactive aggregates and being used in the industry. The dosage of lithium required to control ASR varies with aggregate type and should be established with the specific aggregate-cement combination, prior to use.

Case Study: Precast Concrete Beams with Lithium Nitrate at Rainbow Project (April 2010)

ASTM C1567 Test: ASR tests were performed for precast concrete beams, using 100% Holcim (Types II and III) cement, lithium ASRx 30N (now MasterLife ASR 30), in accordance with ASTM C1567. The average 14-day expansion results, with two different dosages of lithium, are presented in Table 3.8.

HEADWATERS
RESOURCES

Adding Value to Energy™

ASTM C618-05 Testing of
Jim Bridger Fly Ash

Sample Type: 3200-ton **Report Date:** 2/15/2010
Sample Date: 12/11 - 12/29/09 **MTRF ID** 13JB
Sample ID: BR-92-09-R

Chemical Analysis			ASTM Limits Class F	ASTM Limits Class C	ASTM Test Method
Silicon Dioxide (SiO2)	60.63	%			
Aluminum Oxide (Al2O3)	18.19	%			
Iron Oxide (Fe2O3)	4.23	%			
Sum of Constituents	83.05	%	70.0% min	50.0% min	D4326
Sulfur Trioxide (SO3)	0.70	%	5.0% max	5.0% max	D4326
Calcium Oxide (CaO)	6.16	%			D4326
Moisture	0.05	%	3.0% max	3.0% max	C311
Loss on Ignition	0.41	%	6.0% max	6.0% max	C311
Physical Analysis					
Fineness, % retained on #325	21.58	%	34% max	34% max	C311, C430
Strength Activity Index - 7 or 28 day requirement					C311, C109
7 day, % of control	89	%	75% min	75% min	
28 day, % of control	95	%	75% min	75% min	
Water Requirement, % control	93	%	105% max	105% max	
Autoclave Soundness	0.03	%	0.8% max	0.8% max	C311, C151
True Particle Density	2.39				

Headwaters Resources certifies that pursuant to ASTM C618-05 protocol for testing, the test data listed herein was generated by applicable ASTM methods and meets the requirements of ASTM C618-05 for Class F fly ash.

Bobby Bergman
MTRF Manager

Materials Testing & Research Facility
2650 Old State Highway 113
Taylorsville, Georgia 30178
P: 770 684 0102
F: 770 684 5114
www.headwaters.com

Fig. 3.12 Headwaters Resources fly ash chemical and physical analysis results

Table 3.4 ASR expansion ASTM C1567 results for Ranch Pit aggregates

Source	Average 14-day expansion, %	Remarks
Sand/Class C fly ash/cement	0.16	Potentially deleterious
Coarse aggregate/Class C fly ash/cement	0.24	Deleterious
Sand/Class F fly ash/cement	0.05	Innocuous
Coarse aggregate/Class F fly ash/cement	0.04	Innocuous

Table 3.5 ASR expansion ASTM C1260 results for aggregates using low-alkali cement and fly ash

Source	Average 14-day expansion, %		Remarks
	With standard cement only	With low-alkali cement and Class F fly ash	
Ranch Pit sand	0.13	0.05	Innocuous
Ranch Pit coarse aggregate	0.21	0.04	Innocuous

Table 3.6 Type II/V cement mill test results

Requirements	ASTM C150 Type II	Ash Grove Type II/V
Chemical requirements		
SiO_2, %	20.0 min	20.6
Al_2O_3, %	6.0 max	3.7
Fe_2O, %	6.0 max	3.2
CaO, %		63.7
MgO, %	6.0 max	3.2
SO_3, %	3.0 max	1.9
LOI, %	3.0 max	2.4
Insoluble residue, %	0.75 max	0.62
Na_2O equivalent, %	0.60 max	0.45
C_3S, %		59
C_2S, %		15
C_3A, %	8.0 max	4.5
C_4AF, %		10
$C_4AF + 2C_3A$, %	25 max	19
Physical requirements		
Blaine fineness, m^2/kg (air permeability)		402
Minus 325 mesh		99.60
Vicat set time		
Initial, minutes	45 min	130
Final, minutes	375 min	250
Heat of hydration (cal/g) 7 days		92
Air content, %	12 max	8.1
Autoclave expansion, %	0.80 max	0.04
Compressive strength, psi		
1 day		1970
3 days	1450 psi	3180
7 days	2470 psi	4960
28 days		6640

Table 3.7 Potential alkali-silica reactivity, ASTM C1260

Mix #	Mix type	Mix material	14-day expansion, %	28-day expansion, %
Mix 1	ASTM C1260	CA + standard cement	0.16	0.34
Mix 2	ASTM C1260	FA + standard cement	0.19	0.34
Mix 3	ASTM C1260	CA + project cement	0.11	0.30
Mix 4	ASTM C1260	FA + project cement	0.14	0.26
Mix 5	ASTM C1260	CA + project cement + fly ash	0.02	0.02
Mix 6	ASTM C1260	FA+ project cement + fly ash	0.01	0.03

CA coarse aggregate, *FA* fine aggregate

Table 3.8 ASTM C1567 tests with lithium ASRx 30 LN admixture

Cement type (%)	Fly ash (%)	Aggregates	LithiumASRx 30N	Average14-day expansion, %	Remarks
Type II (100)		CA = 59% FA = 41%	0.5 gal/yd³	0.38	Potentially deleterious
Type II (100)		CA = 59% FA = 41%	1.0 gal/yd³	0.03	Innocuous
Type III (100)		CA = 59% FA = 41%	1.14 gal/yd³	0.31	Potentially deleterious
Type III (100)		CA = 59% FA = 41%	2.28 gal/yd³	0.02	Innocuous

Results The ASTM C1567 test results with ASRx 30 LN (now MasterLife ASR 30), at low dosages were not effective. A higher dosage was required for mitigating expansion below 0.1%. Since 2009, CRD C662 expansion test for ASR (with specific combination of concrete mixtures with lithium nitrate) has also provided acceptable results (Fig. 3.13).

The concrete mixture proportions for precast beams, with design compressive strength of 5000 psi, are shown in Table 3.9.

3.8 Other ASR Studies

Reference [12] has published results of ASR expansion on concrete beams and slabs, made from six different concrete using reactive aggregates and different cementitious materials at 708 lb/cy (420 kg/m³) and w/cm ratio of 0.40, monitored for 20 years. The concrete beams 0.6 × 0.6 × 2 m (24 × 24 × 79 in.) and slabs made with high-alkali cement cracked at 5 years: the low-alkali cement cracked after 12 year; the expansion levels with supplementary cementitious materials, 25% GGBFS and 3.8% silica fume with a high-alkali cement, and 50% GGBF slag with a high-alkali cement was such that cracking occurred after 20 years. It concluded:

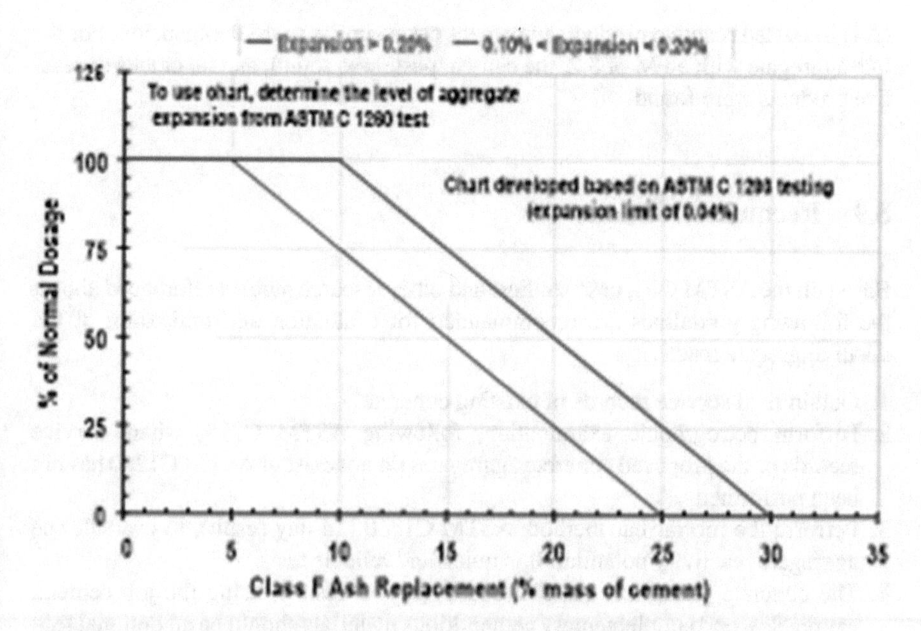

Fig. 3.13 Lithium dosage versus Class F fly ash replacement (percent) (Courtesy of BASF)

Table 3.9 Proportions for precast concrete beams with lithium ASRx 30 LN

Designation	Quantities
Water/cement Ratio	0.39
Cement Type III, lb/cy (kg/m^3)	630
Fly ash, lb/cy (kg/m^3)	
Water, lb/cy (kg/m^3)	231
Sand, SSD, lb/cy (kg/m^3)	1230
#67 stone, SSD, lb/cy(kg/m^3)	1797
AEA, oz/yd	10
Set retarder, oz/yd (ml/m^3)	90
HRWRA, oz/yd (ml/m^3)	117
Lithium ASRx 30 LN, gal/yd	2.28

"The most effective measure at preventing damaging ASR that had also excellent freezing-thawing resistance, was the ternary blend of 25% slag and 3.8% silica fume inter-ground with a high-alkali cement."

Reference [13] studied the effect of fineness modulus (FM) of reactive, metamorphic, fine aggregates, on ASR expansion, using ASTM C1260 accelerated mortar bar test. The fine aggregates FM ranged from 2.4 to 3.25, and the cement had a 0.75% equivalent alkali content. It was found that the ASR expansion decreased as the FM of the aggregate increased. For the fine aggregate with a FM of 3.25 (coarse sand), the 16-day expansion was less than 0.1%. Cross sections of the representative mortar samples were analyzed with scanning electron microscopy (SEM) equipped with energy-dispersive spectroscopy (EDS). The specimen with the smallest FM

(2.4) indicated reactive products in cement paste similar to ASR expansion. For the fine aggregate with a FM of 3.2, the cement paste was sound, and no cracks or reactive products were found.

3.9 Recommendations

Based on the ASTM C33, case studies, and other research studies referenced above, the following guidelines are recommended for evaluation and mitigation of the alkali-aggregate reactivity:

1. Obtain field service records of existing concrete.
2. Perform petrographic examination, following ASTM C295, where service records of the proposed concrete aggregates do not exist or ASTM C1260 has not been performed.
3. Perform the mortar bar method, ASTM C1260 (14-day result), to evaluate the aggregate reactivity potential. It is quick and reliable tests.
4. The concrete prism test, ASTM C1293 (1-year result), using the job cement, aggregates, and supplementary cementitious materials should be an optional test.
5. If ASTM C1260 shows potentially deleterious expansion, perform the accelerated mortar bar method, ASTM C1567, using combination of the cement and supplementary cementitious materials, to limit expansion.
6. Obtain field service condition information of existing concrete using the aggregates in similar structures predicting the effect of environment.
7. Perform a field performance survey of concrete using the same aggregates and cement under the same or more severe conditions.
8. Design concrete mixtures with a low water/cementitious material ratio for water tightness and durability.
9. Use chemical additives, such as lithium-based compounds and liquids, in lieu of supplementary cementitious materials, to control ASR in precast and prestressed concrete, where early compressive strength is required. The dosage of the lithium-based compound should be verified from using ASTM C1260 and CRD C662 tests using the proposed cement, fly ash, and aggregate combinations.

References

1. Stanton TE (1940) Influence of cement and aggregate on concrete expansion. ASCE Proc 66(10)
2. Swenson EG (1957) A reactive aggregate undetected by ASTM tests, bulletin no. 266. ASTM International, West Conshohocken, PA
3. Sims I, Poole AB (eds) (2017) Alkali-aggregate reaction in concrete – a world review. CRC Press, London
4. Long-term operations: literature review of structural implications of alkali-silica reaction. Electric Power Research Institute (EPRI), 2017

5. Concrete manual. United States Bureau of Reclamation, Eighth edition
6. Stark D (1995) Alkali-silica reaction and its effect on concrete, USCOLD Second International Conference, Chattanooga, TN
7. Lenke LR, Mckeen RG, Pallachulla KK, Barringer WL (2000) Mitigation of alkali-silica reactivity in New Mexico, Paper No. 1216, Transport Res Record
8. Alkali-Silica Reaction, Chapter 2. In: Federal highway administration publication, FHWA-RD-03-047, 2003
9. Rangaraju PR, Rao Sompura KR (2005) Influence of cement composition on expansions observed in standard and modified ASTM C1260, Paper 01011203, Transport Res Record, 1914
10. Hicks JL (2007) Mitigation of Alkali-Silica Reaction While using Highly Reactive aggregates with Class C Fly ash and Reduction in Water to Cementitious Ratio, presented at World of Coal Ash
11. Hasan N (2013) Case Studies for Mitigation Alkali-Silica Reaction (ASR) in Concrete, International Conference on Advances in Cement and Concrete Technology in Africa, Johannesburg, SA
12. Doug R, Rogers C, MacDonald CA, Ramlochan T (2013) Twenty year field evaluation of alkali-silica reaction mitigation. ACI Mater J:539–548
13. Yubin J, Chi-Sub J (2010) Effect of fineness modulus of reactive aggregate on alkali-silica reaction. Int J Concr Struct Mater 4(2):119–125

Chapter 4
Concrete Mixing Placing and Curing

4.1 General

All concrete used in the work should be produced at a single batching plant and material sources to provide to uniformity and consistency of delivered concrete. All aspects of the batching, mixing, and transportation (equipment and procedures) should be established ahead of construction. An independent backup plant is recommended to assure reliable production of concrete.

Concrete construction should be in accordance with the applicable requirements of the references listed herein.

Concrete formwork should be designed in accordance with ACI 347.

Slump: The consistency of the delivered concrete should be monitored by slump tests.

Admixtures should be used only when required by constructability or service or exposure condition or to enhance durability (such as freeze-thaw, sulfate attack, etc.). Use of admixtures would not relax cold weather placement requirements.

Refer to Chaps. 1 and 2 for concrete materials and concrete mixture proportions design. For quality control, refer to Chap. 9.

ASTM C94 provides standard specifications for manufacture and delivery of fresh concrete. Standards for the Concrete Plant Manufacturers Bureau (CPMB) and Truck mixer Manufacturers Bureau (TMMB) can be found on the National Ready Mixed Concrete Association's (NRMCA) website at http://222.nrmca.org.

Refer to ACI 304R, ACI 308, and ACI 309 for additional information on mixing, transporting, and placing of concrete. ACI 309 and ACI 308 provide additional information on consolidation of concrete and curing of concrete, respectively.

N. Hasan, *Durability and Sustainability of Concrete*,
https://doi.org/10.1007/978-3-030-51573-7_4

4.2 Concrete Batch Plant

Concrete should be supplied from an automatic ready-mixed concrete plant, including a backup plant, located on site or offsite. The plant(s), including scales, meters, handling methods, and operating personnel, should be certified in accordance with be applicable State Department of Transportation (DOT) and applicable NRMCA standards.

Typical batch plant specification for a 150–200 cy/hr.

- A nominal rated output capacity of 200 cy/hr. based on a 10-cy batch size, 120-second mix time, and 30-second mixer dump rate into open top agitator trucks.
- A tilting drum mixer – 12 cy CPMB rated capacity, capable of mixing 12 cy of nominal 1-inch slump concrete and 12 cy of 12-inch slump concrete without spillage. Mixers should be mounted above collecting cones with a minimum 13′ clearance below the discharge outlet.
- Collecting cone shall be capable of loading mixed concrete into agitator trucks or conveyor feed hopper.
- Four aggregates – with nominal 150 cy (CPMB rated) overhead storage in four separate compartments, each with high-level indicators.
- Individual aggregate weigh batchers with separate scales for four aggregates. The sand batcher should be capable of weighing mixes with up to 65% sand content.
- Means to empty the aggregate storage bins, with either a reversing collecting conveyor or a mixer bypass chute. Two overhead silos with a combination of three separate compartments for the following materials: Portland cement, fly ash, and GGBFS. Each silo to be nominal 1000+ bbl. capacity. Include individual dust collectors for each silo compartment. Include dual 5″ fill pipes for each silo compartment.
- Two separate cement individual weigh batchers with one for weighing of cement (oversized capacity for cumulative batching of fly ash and cement) and individual weighing of GGBFS. Each weigh batcher should include suitable sized filter vent and air-operated vibrator. Include provisions for possible future addition.
- 48″ batch transfer conveyor.
- Minimum 30 HP rotary screw air compressor with 120-gallon air-receiving tank.
- Nominal 4″ water meter with strainer, suitable for hot and cold water service. Include a nominal 2″ tempering water meter with strainer. Include all necessary control valves, necessary connecting hoses and fittings, and controls.
- Moisture meters in the sand bin, along with one in the 3/4″ aggregate bin and the 1–1/2″ aggregate bin. The fine aggregate moisture meter should meet the requirements of Corps of Engineers (COE) Standard CRD-C 143.

- Live bottom ice batcher and day tank each sized for nominal 75 cubic foot capacity. Ice weigh batcher should be variable speed discharge.
- Fully automatic electronic plant controls with full recording and minimum of 50 preset mixes – full interlocks, control of six admix dispensers, IP to metric interchangeability, and manual backup panel.
- Plant to meet all applicable OSHA standards.
- Equipped with necessary access ladders, service platforms, and handrails.
- Two sets of standard test weights – 1000 lbs. minimum, 50 lbs. each.
- Plant to be painted one shop coat of primer and one shop coat of medium gray color paint minimum.
- Dust collection equipment that meets the requirements of EPA and OSHA and local regulations.
- Control house, minimum size 8′ × 20′ with lights, air conditioning, and heat, including support structure for mounting at the mixer deck level.
- Structural design – meet the requirements of the latest UBC and/or local governmental agencies. Plant structure shall be designed to meet UBC requirements for the designated site.

4.3 Batching and Mixing of Materials

Batching Equipment: The quantities of cement, pozzolanic materials, and aggregates entering each batch of concrete should be determined by automatic weighing measurements by mass. The batching system should be capable of proportioning any of the designated concrete mixtures. The quantity of water and admixtures may be determined by weighing or volumetric measurement. Batchers shall conform to the Concrete Plant Standards.

Discharge of supplementary cementitious materials (SCM) such as fly ash into the cement hopper should be only after the addition of the Portland cement. Store fly ash in such a manner that allows ready access for inspection and sampling. Protect fly ash from contamination and moisture. If any fly ash shows evidence of contamination or unsuitability, it should be rejected and removed from the site.

The weighing equipment for aggregates should be capable of ready adjustment both to compensate for variation in moisture content of the aggregates and for changing concrete mix proportions. The concrete material and water/ice should meet the batching tolerances of ASTM C94 (±1% for cementitious materials and water, ±2% for aggregates, ±3% for admixtures).

Prior to production, the ready-mixed plant and weighing equipment should be inspected for accuracy in accordance with the NRMCA standards.

Case Study: San Roque Project, Philippines

San Roque: 160 m^3/hour concrete batch plant

An on-site concrete plant was considered essential to meet the fast-track construction schedule and concrete placement requirement for the project. A 210 yd^3 (160 m^3) per hour plant capacity was considered adequate to meet the concrete placement schedule, assuming a two-shift daily operation, 6 days a week. The new plant was ordered approximately 1 year ahead of the actual concrete production, which allowed reasonable time for fabrication, delivery, and erection and trial mixes testing on site.

The fully automatic, concrete plant was equipped with the following:

- 5-compartment aggregate bin with 288 yd^3 (225 m^3) capacity and a minimum of four individual batchers for weighing aggregates
- A cementitious bin with 1000 barrels (170 metric tons) rated capacity, separated 1/3 and 2/3 by a partition
- 850 metric tons of additional cementitious storage silos
- A cumulative weigh batcher for cement and fly ash
- A silo and weigh batcher for silica fume
- 70 metric tons ground storage silo for condensed silica fume
- An automatic ice flake plant with 55 metric ton per 24 hours capacity
- A prefabricated 100 ton insulated ice storage building
- An admixture dispensing system capable of distributing a minimum of two separate admixtures in a batch
- Two 10 yd^3 (7.75 m^3) tilting type stationary mixers suitable for 3-inch (75-mm) coarse aggregate for low slump concrete
- A weatherproof and air-conditioned control trailer
- A central dust collection system

The ice plant was provided to lower the concrete batch temperature to 68 °F (20 °C) maximum for massive concrete placements. Given the high relative humidity at the site, water spraying the aggregate stockpiles was not a considered option; however, the 3-inch (38-mm) coarse aggregate bin was cooled with a 50 °F air for lowering the stockpile temperature.

The batch plant was erected in January 2000. The concrete trial mixes program was conducted in February through April 2000. Each trial mix (in 2 cubic meters concrete batches) was tested for workability, slump, air content, and compressive strength. The regular production of concrete commenced in June 2000 and continued for 30 months. The average daily production varied from 500 to 600 cubic meters (600–784 cubic yards per day). The amount of concrete placed, by structures and class of concrete, is shown in Table 4.1.

Refer to Table 2.15 for concrete mixture designs (Figs. 4.1, 4.2, and 4.3).

Table 4.1 Concrete quantities (in cubic meters)

Structure	Concrete quantities m^3 (yd^3)
Spillway	186,200 (243,300)
Spillway dental	37,200 (48,660)
Power tunnel	59,000 (77,170)
Powerhouse	24,000 (31,400)
Low-level outlet tunnel	28,200 (36,620)
Tailrace and switchyard	18,400 (24070)
Diversion tunnels/plugs	29,800 (38,980)
Galleries	9700 (12,790)
Total	*392,500 (513,370)*

Fig. 4.1 Concrete catch plant

Fig. 4.2 12-cy central mixer at Chickamauga, TN

Fig. 4.3 A close-up photo of the 12-Cy central mixer

Case Study: Rainbow Redevelopment Project, Great Falls, MT

A central mix concrete plant, rated at 250 cy/hour, manufactured by Erie Strayer Company, was erected on-site for central mixing of concrete. The plant was equipped with the following:

80 cy capacity aggregate bins with three compartments
12 cy aggregate batcher mounted on a load cell scale system
48-inch-wide batch transfer conveyor with 60 HP electric motor
Cement silo CS@-715, split 1/3–2/3 compartments (71 cy and 35 cy capacity)
Cement batcher CB2-12 mounted on a load cell scale system and air-operated butterfly discharge valve
Aeration blower 4.5 HP capacity mounted on cement batch platform
12 cy tilting mixer with 75 HP mixer drum drives motors and gears, 50 HP hydraulic pump drive motor, tilt speed control valve
Water meter, control valves
Plant control panel
Dust collection system
Command Alkon Eagle Batching System for automatic batching

In addition, a Worthington central mix concrete plant, rated at 140 cy/hour, located in Great Falls, was available to meet the concrete production for the mass concrete placements. This plant was equipped with a 6-cy drum mixer and Command Alkon batching computer.

The concrete from the plants was conveyed in McNeilus Bridgemasters transit mixers with 11 cy capacity. The travel times for the site plant and the Great Falls plant were 3 minutes and 20 minutes, respectively.

Centralized Batch Mixers Centralized batching and mixing plants should have a stationary mixer with a rated capacity suitable to the concrete placement rates. For a large project, requiring placements of 100–250 cubic yards, the central mixer should be 10-cubic yard minimum capacity.

Mixers in centralized batching and mixing plants should be arranged so that mixing action can be observed from a location convenient to the mixing-plant operator's station.

Mixer uniformity tests should be made in accordance with ASTM C94 at the time of initial start-up. The tests should be made with full size mixer load and for the coarsest size aggregate. Mixing time should be established by the mixer uniformity tests, prior to production. Unless verified by uniformity test, concrete ingredients should be mixed for not less than 1.5 minutes after all the ingredients, except the full quantity of water, are in the mixer.

Each stationary mixer should be equipped with a mechanically operated timing and signaling device that indicates the completion of the required mixing period and count the batches.

Concrete Association (NRMCA) Plant Certification Check List and a Certificate of Conformance should be provided. Scales for individual batchers should be retested, adjusted as required, and recertified on a regular basis.

Mixing Methods There are several ways in which concrete from central batching plants can be handled.

Central-mixed concrete is completely mixed in a stationary mixer at the plant and transported to the point of delivery in truck mixer or agitator operating at agitating speed.

Shrink-mixed concrete is first partially mixed in stationary mixer and then completely mixed in a truck mixer during the time the concrete is being transported to the job site.

Truck-mixed concrete is completely mixed in a truck for 70–100 revolutions at the mixing speed designated by the manufacturer.

Truck mixers should be operated to ensure that the discharged concrete uniformly complies with the requirements. Truck mixers should be equipped with approved revolution counters for which the number of revolutions of the drum or blades may readily be verified. The water-tank system of the truck should be equipped with gauges that provide accurate indication of the tank contents.

Each batch of concrete should be mixed in a truck mixer for not less than 70 revolutions of the drum or blades and at the rate of rotation designated as mixing speed by the manufacturer of the equipment. Additional mixing, if any, should be at the speed designated as the agitating speed by the manufacturer of the equipment. All materials, including mixing water but excluding high-range water reducers, should be in the mixer drum before activating the revolution counter for determining the number of revolutions of mixing.

Transit mixing of concrete should comply with ASTM C94.

Dispensing of Admixtures Dispensers for admixtures should have the capacity of the full quantity of the properly diluted solution required for each batch. They should be maintained in a clean and freely operating condition. Admixtures should be added in accordance with manufacturer's recommendations including the dosage rates. The admixture should be discharged into the batch by flowing automatically

Fig. 4.4 Typical aggregate stockpile at the batch plant

Fig. 4.5 San Roque aggregate processing

and uniformly into the stream of mixing water from beginning to end of its flow into the mixture. Dosing of admixtures should be accurate to within five percent. Equipment for measurement should provide visual confirmation of the accuracy of the measurement for each batch.

High-range water-reducing admixture should be dispensed at the plant or at the jobsite, with slump not exceeding the working limit. The high-range water-reducing admixture should be accurately measured and pressure-injected into the mixture as a single dose by a qualified technician. The field dispensing system should conform to the same requirements as a plant system, and a standby dispensing system should be provided and tested prior to each day's operation of the primary system. After the addition of the high-range water reducer, the concrete should be mixed at mixing speed for a minimum of 5 minutes.

Field addition of admixture and/or water must be such that the design w/cm ratio is not exceeded.

4.4 Transporting Concrete to Site

Transporting equipment should be designed and operated so that it does not cause segregation or loss of material. The capacity of the transporting equipment should not exceed the batch plant capacity to ensure a continuous supply of concrete.

Transport and deliver concrete in equipment conforming to ASTM C94. Only agitating equipment should be used to transport concrete on roads or rails. Transit mixing and long-distance transportation of mixed concrete would not be permitted in any case where control of measurement of cement, aggregates, water, concrete temperature, or any other operation affects the quality of concrete so that it fails to comply with the slump and air content requirements. Discharge of concrete should normally be completed no later than 90 minutes after the introduction of mixing water to cement or 300 revolutions of drum and 60 minutes during hot weather. The discharge of concrete may be delayed beyond these limits with the use of an approved retarding admixture and subject to site testing to determine its effectiveness.

Prior to use, truck mixers and agitators should be tested for compliance with the concrete uniformity requirements of Appendix A1 of ASTM C94. All mechanical details of the mixer, such as water measuring and discharge apparatus, condition of the blades, speed of rotation, general mechanical condition of the unit, and clearance of the drum, should be checked before use.

Non-agitating equipment should not be used for transporting concrete.

4.5 Handling and Placing

Table 4.2 summarizes the most common methods and equipment for handling and placing concrete [1].

Table 4.2 Equipment for handling and placing concrete

Equipment	Type and range	Remarks
Belt conveyors (Telebelts)	For conveying concrete horizontally or to higher or lower level Types: portable, feeder, or spreading conveyors Discharge concrete up to 35 ft height. Placement rates of over 100 yd^3/hr. (76 m^3/hr.)	Can place large volume, with adjustable speed and reach, where access is limited. Concrete needs to be protected against inclement weather No limitation on NMS aggregate
Pumps	For conveying concrete directly from truck discharge point to formwork, placement booms support a 4/5-inch pipeline, with reach of 72–175 ft and varying rated capacity	Can place large volume, with adjustable speed and reach, where access is limited Require pumpability tests to establish slump and air content of delivered concrete Limitations on NMS aggregate (1–1/2 inch)
Buckets	For conveying concrete directly from truck discharge point to formwork	Clean discharge; however, placement rates are based on bucket size. Maintains uniformity of concrete delivered. No limitation on NMS aggregate
Tremie	For placing concrete underwater	Can be used to funnel concrete directly into the foundation underwater. Need advance planning, mixture design, and placing procedure
Drop chutes	Used for placing in vertical form. Tubes are flexible, made from plastic or metal (elephant truck), and can be used for over hundreds of feet with the use of a remix chamber	Concrete is placed directly into form without segregation
Cranes and buckets	For concrete placement above or below ground	Can also handle reinforcing steel, forms, and construction items

Pumping Equipment The pumps should be squeeze-pressure type. Pneumatic placing equipment should not be used. The pump lines should be rigid steel pipe. The minimum diameter of the flexible hose should be at least three times the NMS aggregate in concrete to be pumped, but not less than four (4) inches. Aluminum pipe should never be used. Pumping equipment and hoses (conduits) that are not functioning properly should be replaced.

Pumping may result in a loss of slump and entrained air between the truck discharge and the final placement. The type and operation of concrete pumps and pneumatic or other types of placing equipment should be selected to facilitate concrete placement with minimum slump loss and entrained air loss. The equipment used and its method of operation should allow placement of concrete into the forms without high velocity. Placing equipment should be operated only by qualified operators.

All pump priming slurry should be discarded before placement. Discarding the priming slurry may require that several cubic yards of concrete be discharged through the pumping system and discarded. Use a reserve pump or other backup equipment at the site.

The pumping system consists of a hopper into which concrete is discharged from the mixer, a concrete pump, and the pipes through which the concrete is pumped (Figs. 4.6, 4.7, and 4.8).

Belt Conveyors (Telebelts) The belt conveyors for transporting concrete should be designed to ensure a uniform flow of concrete from point of origin to final placement without segregation of ingredients or loss of fines or mortar. Belt width should be 16 inches minimum for NMSA of 3 inches and under. Belt speed should be 300–600 feet per minute range and preferably have a length no greater than 300 feet.

Fig. 4.6 San Roque spillway wall construction using a boom pump

Fig. 4.7 San Roque spillway flip bucket construction using a boom pump

Fig. 4.8 A concrete bucket discharging concrete

Discharge baffles should be provided at transfer points. If concrete needs protection from sun or rain, a cover over the belt should be provided. All parts of the conveyor must be accessible for inspection.

Belt conveyors, such as Putzmeister 110G, can place concrete up to 360 yd^3/hr. (275 m^3/hr.).

Chutes are used for placing concrete directly from a truck mixer, agitator, or non-agitating truck.

Transfer hoppers should be capable of receiving concrete directly from delivery vehicles and should be conical in design. It should be equipped with a hydraulically operated gate and means of external vibration to effect complete discharge (Figs. 4.9 and 4.10 and Table 4.3).

Table 4.3 Rainbow powerhouse project equipment for placing concrete

Pumps	42X Boom Putzmeister
Telebelts	TB 110G Putzmeister
Buckets	Cl-100-200

Fig. 4.9 Putzmeister 110 G Telescopic Telebelt

Fig. 4.10 Putzmeister 42 X Boom Pump with 5-inch slickline vertical reach of 137 ft

4.6 Placing Concrete

General Immediately before placing reinforcing steel and concrete, all surfaces of foundations against which the concrete would be placed should be free from standing water, mud, oil, harmful coatings, and debris. Rock surfaces should be washed free of all loose material with air-water jets.

Earth surfaces, after they have been trimmed and compacted, should be thoroughly moistened, but not muddied, and should be kept moist by frequent sprinkling as necessary until the concrete is placed against it. A working unreinforced slab, or a mud mat, is recommended over an earth subbase. It provides a solid working surface, support of reinforcing steel, easy clean-up prior to placement of concrete, and schedule effectiveness.

All surfaces of forms, reinforcing steel, and embedded items that have been encrusted with mortar or grout from previously placed concrete should be cleaned of all such mortar and grout before the surrounding or adjacent concrete is placed.

Hardened concrete surfaces to which new concrete is expected to bond or where concrete placement has been interrupted so that the new concrete cannot be placed monolithically are defined as construction joints.

Where new concrete is to be placed against hardened concrete, the surfaces should be kept thoroughly wet during the 24-hour period immediately prior to the placement of the new concrete. This wetting should be accomplished by continuous sprinkling or by covering with burlap that is kept wet. All existing projecting steel bars to be embedded in new concrete should be cleaned.

Bonding compounds should be used only where surfaces cannot be roughened. Bonding compounds should be applied in accordance with the manufacturer's printed instructions, except that the compound should be applied by brush. Care should be taken to keep the compound off the surfaces of the grooves in which sealant is to be placed. Bond-breaker should be used only where specified.

Concrete should not generally be placed until all water entering the space to be filled with concrete has been properly cut off or diverted by pipes or other means. Where concrete needs be deposited under water, procedures for tremie placement developed, as described later in this chapter. Do not allow still water to rise on any concrete until the concrete attains its initial set. Water should not be permitted to flow over the surface of any concrete in any manner that would injure the surface of the concrete.

When concrete slabs or other exposed concrete are placed under warm, dry, or windy conditions, protect the concrete surfaces from drying by using windbreaks, shading, or fogging. Proper dust control should be provided during such placement. If these conditions are not satisfactorily met, concrete should not be placed.

4.7 Placement Guidelines

1. Placement plans and sequence of concrete placement should be submitted and approved, prior to use.
2. Embedded items: Before concrete is placed, all reinforcement, waterstops, expansion board, anchor bolts, edge angles, sleeves, drains, and other materials and equipment should be accurately and properly placed in accordance with approved shop drawings or as directed. They should be secured against displacement during placement of concrete. All embedded items should have a minimum specified cover to the nearest outer surface of concrete.
3. Cleaning: All ferrous metalwork to be embedded in concrete should be free of rust, loose scale, dirt, oil, grease, and other foreign substances, before concrete is placed. Cleaning should be done by metal brushes, scrapers, chisels, sandblasting, or grit blasting.
4. Inspection: All formwork, reinforcing steel, waterstops, and other embedment should be inspected and approved prior to placing of concrete. Any defective or incomplete work discovered by inspection should be corrected soon enough to allow re-inspection and further correction if necessary.
5. Construction joints: The order of placing concrete and the number, type, position, and design of construction joints should be as shown in the drawings and approved. In formed concrete, a transverse construction joint with a roughened face should be provided at the end of any continuous placement. All horizontal construction joints should be made straight and level. All vertical construction joints should be plumb. When a cold joint is formed, the procedure used for placing of concrete should be the same as for a construction joint. The surfaces of construction joints should be cleaned and roughened to a 1/4-inch amplitude.
6. Concrete should be placed as soon as possible after leaving the mixer and without segregation. Concrete should not be dropped from a height that causes segregation. Concrete should not be allowed to leave an accumulation of mortar on the form surfaces above the placed concrete. Hoppers and vertical ducts should be used to deposit concrete as nearly as practical in its final position and without segregation. The free unconfined fall of concrete and the horizontal movement of concrete in the forms should not exceed five (5) feet, except for SCC (refer to Chap. 6).
7. Do not place concrete on frozen or ice-coated ground or subgrade; against ice-coated forms, reinforcing steel, structural steel, conduits, or construction joints; where vibrations from nearby work may harm the concrete's initial set or strength; or during rain.

8. Except as intercepted by joints, all concrete should be placed in continuous, normally horizontal layers, the depths of which generally should be range of 12–20 inches. Lesser depths may be required to ensure that each new layer can be made monolithic by penetration of the vibrators. Concrete placement should begin at the lowest point in each section of concrete to be placed. Ensure that the means and methods used for concrete placement do not cause misalignment of the embedded items.
9. Placement interval: Each lift of concrete should be allowed to set for at least 72 hours before the start of a subsequent or adjacent placement.
10. Placing concrete through reinforcing steel: When placing concrete through reinforcing steel, care should be taken to prevent segregation of the aggregate. The initial placement should be slow until a puddle develops to above the reinforcing mat.
11. Temperature: The temperature of concrete as placed should not be more than 90° F, except the placing temperature for mass concrete should not be more than 70° F. During hot weather, follow the recommendations in ACI 305R in order to keep the concrete temperature below the specified maximum. Any wetting of aggregates for cooling should be performed sufficiently in advance of delivery into the batching plant bins and in such manner that the batched material would have a uniform and stable moisture content.
12. Concrete should not be mixed or placed when the ambient air temperature is 40° F or less and dropping and the possibility of freezing is indicated by the US National Weather Service frost warnings, unless arrangements for protection of concrete have been made.
13. Concrete production, delivery, and protection during cold weather should be in accordance with ACI 306R. Insulating blankets and external heat sources should be used, as necessary, to maintain the external concrete temperature within 35° F of the internal concrete temperature. Thermocouples or sensors and monitoring devices should be embedded in the concrete to monitor internal concrete temperature. Thermocouples and sensors should be placed a minimum of 2 inches from the exposed surface. Embedded sensors to estimate the in-place concrete strength, using a strength-maturity relationship, should be in accordance with ASTM C1074.

Consistency The quantity of water entering into a batch of concrete should be just sufficient, with a normal mixing period, to produce a concrete that can be placed without segregation and that can be consolidated to provide the desired density, impermeability, and smoothness of surface.

The slump of the concrete immediately prior to placing should be in accordance with the limits in Table 4.4, measured in inches.

The slump should be measured in accordance with ASTM C143.

When a Type F or G high-range water-reducing admixture is allowed in concrete to facilitate placement, the slump of concrete should be 4 inches maximum, before the admixture is added.

Table 4.4 Slump requirements

Strength class (psi	Title	Working limit	Inadvertency margin	Rejection limit
3000 and 4000	Concrete with water-reducing and set-retarding (Type A to D) admixtures	4	1	5
4000 and 5000	Concrete with high-range water-reducing and set-retarding (Type F and G) admixtures	8	1	9
3000	Mass concrete	3	1	4

Note: Concrete and mortar should generally not be retempered. Any concrete or mortar that has stiffened so that proper placement and consolidation cannot be performed should be discarded

Slip-Forming

This is a continuous process of placing and compacting low workability concrete.

Both horizontal and vertical slip forming is possible; vertical slip forming is slower, requiring formwork until sufficient strength has been gained to support the new concrete and the form above.

It is used in the construction of silos, chimneys, monolithic tunnel linings, and high-rise construction.

Horizontal slip-forming can be used for long linear elements such as highway safety barrier.

This arrangement will decrease the construction time as there is no need for stripping and re-setting.

Slip forms are in general built up of sections which could be raised or lowered by the help of jacks or screws.

This is done by means of arranging guiding rails as which steel rods or pipes, well braced together to carry the forms continuously in the required direction. Surfaces of forms (internal) are to be oiled before concreting to prevent sticking or dragging of concrete during moving the forms.

Case Studies for Slip-Forming [2]

In the late 1970s, slip-forming of the heavily reinforced concrete cylindrical shield walls at several nuclear plant sites, in Louisiana (Waterford Unit 3, Fig. 4.11), Florida (St. Lucie Unit 2), and Washington (WPPSS Units 3 and 5), was successfully implemented with savings in cost and schedule. It proved that, with proper planning, design, and coordination, the slip-form construction technique can be successfully applied to large and heavily reinforced concrete structures. The shield walls for WPPSS were designed in compliance with seismic requirements which resulted in the need for reinforcing steel averaging 550 pounds per cubic yard (326 kg/m^3). A 25-ft (7.6-m)-high, three-deck moving platform was designed to permit easy installation of the reinforcing steel and embedment, to block out, and to facilitate concrete placement and finishing. The rates of placement varied from 6 to 2 inches per hour (Ref. [2]).

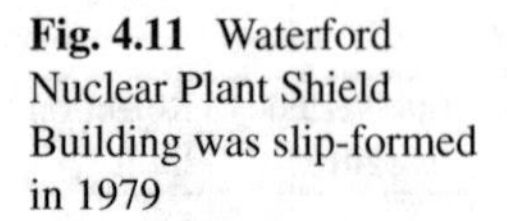

Fig. 4.11 Waterford Nuclear Plant Shield Building was slip-formed in 1979

At San Roque, the 4-meter-thick piers at the spillway ogee structure were slip-formed by L and M Maxon Co, Inc. using a 4000 psi concrete mixture A2-11E with 1–1/2 inch coarse aggregate (refer to Table 2.15). The average concrete volume in the 4-m-wide by 30-m-long by 30.2-m-high pier was 4700 m^3. The concrete was placed continuously at an average rate of 6 inches per hour vertically with savings in schedule, as shown in Fig. 4.12.

Tremie Concrete Placement

General This is particularly suited for deep forms, where compaction by the usual methods is not possible and for underwater concreting.

High workability concrete is fed by gravity through a vertical pipe which is gradually raised. The method is advantageous where minimal surface disturbance is needed, especially when a concrete water interface exists. The mix should be cohesive, without segregation or bleeding, and usually has a high cement content, a high proportion of fines (40–50%), and admixtures for workability.

The slump of such a tremie mixture, including HRWRA, should range between 6 and 10 inches to mitigate segregation and promote self-leveling. For successful tremie placement, a steady uninterrupted flow of concrete through the tremie pipe should be maintained.

Fig. 4.12 Piers at San Roque Intake Structure were slip-formed

4.8 Placement Guidelines

1. Tremie pipe should be rigid, heavy gage steel and watertight, and sufficiently large to permit a free flow of concrete. Minimum diameter of the tremie should be 8–10 inches. A tremie plug (solid go-devil) or a watertight plate is required prior to initiating placement.
2. Spacing of tremie pipes should be 15–20 feet apart maximum.
3. A stable, support platform is required to support the tremie during placement.
4. Concrete should be designed to be self-compacting, with a viscosity-modifying admixture, with a slump of 6–10 inches or slump flow of 22 inches (ASTM C1611), and tested for segregation prior to use.
5. Concrete may be delivered to the tremie using a concrete pump; however, the pump line may not be attached to the tremie pipe.
6. The tremie bottom should be placed a minimum 6 inches from the bottom to allow water and concrete to escape. Tremie should have markings at 1-ft interval to verify volume of placement and raising points.
7. Insert a go-devil tremie plug.
8. Place a hopper or funnel above the tremie pipe. The hopper should have adequate capacity to for uninterrupted concrete placement. A 4 ft × 4 ft hopper with 8-inch outlet is recommended for large tremie placements.

9. Tremies using an end plate should be filled with concrete being raised a maximum of 6 inches to initiate flow. Tremies using a go-devil should be lifted a maximum 6 inches to allow water to escape. Add concrete to the tremie slowly to force the go-devil downward. Then lift the tremie enough to allow the go-devil escape.
10. Add concrete continuously to assure pipe is kept full of concrete.
11. Keep the tremie embedded in the fresh concrete minimum 3 feet deep.
12. Calculate the concrete placement per tremie location. The tremie pipe placement should limit the horizontal flow of concrete to 15 feet maximum. For example, if a placement depth is 8 ft, assuming a concrete spread of 12 feet, it will require 4 cy/ft depth or 32 cy for the full height.
13. Monitor placement rate with drop lines to verify actual depth. If there is a delay in concrete placement greater than 30 minutes, tremie pipe should be pulled and resealed.
14. Do not move the tremie horizontally during a placing operation.
15. If a loss of seal occurs, evident by increased flow rate in the tremie, follow the initial reseal procedure prior to continuing.

4.9 Cold Weather Placement

Cold weather is defined in ACI 306R as a period when, for than three consecutive days, the average daily temperature is less than 40° F (4.4° C) and the ambient temperature is not greater than 50° F (10° C) for more than one-half day of any 24-hour period. The average daily temperature is the average of the highest and the lowest temperatures occurring during the period.

Cold weather provisions, in accordance with ACI 306R, include the following:

1. An accelerating admixture, conforming to ASTM C494, Type C, is added to concrete, during cold weather only, to reduce the setting time and strength gain.
2. Water is heated to 140F. To facilitate mixing of concrete, 3/4 of the water was added to the drum before the addition of coarse aggregate.
3. The temperature of concrete as delivered is 45F minimum.
4. A protective enclosure with external heat source is installed prior to placement of concrete.

Sensors are embedded in concrete to monitor concrete temperatures during curing. In addition, thermal insulating blankets are used to protect the concrete surfaces during the curing period. The monitoring devices were read every 30 minutes. The protection is continued until the differential concrete temperature between the surface and the interior of the concrete is less than 35F.

During cold weather, concrete placement temperature should meet the recommended temperatures noted in Table 4.5. Mixing water should be heated if necessary, to maintain the recommended concrete placement temperatures. All surfaces in contact with concrete should be maintained at 32F minimum and no higher than 25F higher than the placement temperatures.

Table 4.5 Cold weather concrete placement recommendations

Member thickness	Concrete temperature °F (°C)	Concrete curing temperature, °F (°C)	Minimum curing period, days	Maximum 24-hour concrete temperature drop at end of curing °F (°C)
Less than 1 ft.	55–75 (12–24)	55 (12)	7	50 (10)
>1<3 feet	50–70 (10–21)	55 (12)	7	40 (10)
3–6 feet	45–65 (7–18)	50 (10)	7	25 (−4)
>6 feet	40–60 (10–15)	50 (10)	7	25 (−4)

Concrete production, delivery, and protection during cold weather should be in accordance with ACI 306R. Insulating blankets and external heat sources should be used, as necessary, to maintain the external concrete temperature within 35 °F of the internal concrete temperature. Thermocouples or acceptable sensors and monitoring devices should be embedded in the concrete to monitor internal concrete temperature. Thermocouples and sensors should be placed a minimum of 2 inches from the exposed surface. Embedded sensors to estimate the in-place concrete strength, using a strength-maturity relationship, should be in accordance with ASTM C1074.

4.10 Hot Weather Placement

Hot weather is defined in ACI 305R as any combination of the conditions that impair the quality of freshly mixed or hardened concrete by accelerating the rate of moisture loss and rate of cement hydration concrete due to high ambient temperature, high concrete placement temperature, low relative humidity, wind speed, and solar radiation. During hot weather mitigation measures should be developed including use of a retarder in concrete, limiting placement temperatures and placing concrete within 60 minutes after mixing. Soon after finishing, begin curing with protect the exposed surfaces (see curing below) with wet blankets to minimize surface drying. Figure 4.13 shows a relationship between wind velocity, relative humidity, and concrete temperature to estimate the rate of evaporation from the exposed concrete surface.

4.11 Consolidation of Concrete

Equipment and procedures for the consolidation of conventional concrete should conform to ACI 309R. Concrete vibrators should not be used to transport concrete inside the forms. Vibrators should be inserted in overlapping pattern, vertically, and penetrate into the preceding layer, before concrete reaches initial set. The duration should be sufficient, generally 5–15 seconds, to consolidate the concrete without causing segregation. The vibrator should then be withdrawn at a slow rate to allow entrapped air to escape. Effectively and quickly. Vibrators should not be used for shifting or dragging concrete.

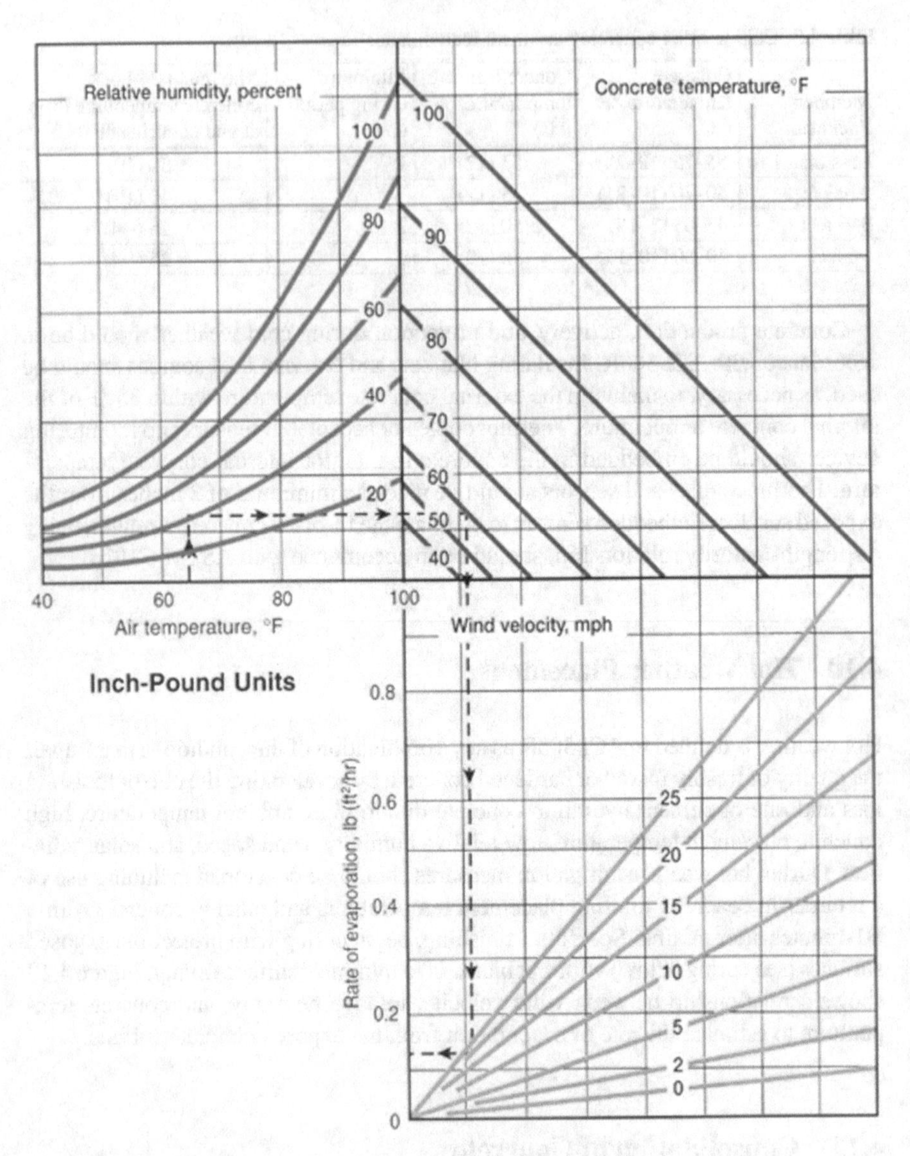

Fig. 4.13 Effect of concrete and air temperatures, relative humidity, and wind velocity on the rate of evaporation of surface moisture (Courtesy of PCA, Ref. [1])

Sufficient number of vibrators and backup vibrators should be provided to ensure uninterrupted consolidation.

Consolidation: Consolidate all structural and mass concrete, with mechanical vibrators having a frequency and an amplitude adequate to consolidate the concrete. At least one standby vibrator in operable condition should be at the placement site prior to placing the concrete. The forms shall allow visual observation of the concrete, and the vibrator operator should be able to see the concrete being consolidated.

Consolidated concrete should be judged satisfactory, when large aggregate is well embedded, a thin layer of glistening mortar lies on the surface, air bubbles are eliminated, and the concrete has generally levelled off.

4.12 Concrete Finishes

General Finishing of concrete surfaces should be performed in accordance with ACI 301 only by skilled workers. Concrete surfaces should be inspected to determine that surface irregularities are within the specified limits.

Formed Surfaces: All finished formed surfaces should conform accurately to the shape, alignment, grades, and sections shown on the drawings. The surfaces should be free from fins, bulges, ridges, offsets, honeycombing, or roughness of any kind and shall present a finished, smooth, continuous, hard surface. Precautions should be taken to ensure that any concrete surfaces that will be exposed to flowing water will have the smooth and even texture produced by the proper use of the best forms and the best and most modern means of concrete placement. Concrete surfaces that will not be exposed to flowing water should be finished in a workmanlike manner and should be free from significant defects, but may not require the same workmanship specified for the surfaces that will be exposed to flowing water. External corners should be rounded or beveled, using molding strips or with suitable molding or finishing tools

4.13 Type of Finishes

Finish F1: This finish concerns surfaces that will be hidden from sight, or if exposed, the aesthetic appearance is not important. No special treatment is required except for the repair of defective concrete.

Finish F2: This finish concerns surface exposed to sight, but without aesthetic requirements. Surface irregularities should not exceed 1/2 inch if they are abrupt and 3/4 inch if they are gradual, measured with a 10-ft template.

Finish F3: This finish concerns surfaces, exposed to sight, where contour must be accurate and without irregularities. Surface irregularities should not exceed 1/4 inch if abrupt and 1/2 inch if gradual, measured with a 10-ft template (Figs. 4.15, 4.16, and 4.17).

Finish F4: This finish concerns surfaces exposed to water where smoothness is of utmost importance from hydraulic point of view. Surface irregularities should not exceed 1/8 inch if abrupt and 1/4 inch if gradual measured with a 10-ft template (Figs. 4.14 and 4.18).

Fig. 4.14 San Roque 8.2-m-diameter tunnel under construction. Form transporter in the background

Fig. 4.15 San Roque powerhouse turbine generator Stage 1 concrete

Fig. 4.16 San Roque powerhouse turbine generator second stage concrete

Fig. 4.17 San Roque concrete placement in slab at El 109.5 m. (Bridge crane is at El 117.75 m)

Unformed Surfaces All exposed unformed surfaces should have a wood float finish. Invert slabs, floors, and flat surfaces should also be generally sloped to drain at a slope of not less than one-eighth (1/8) inch per foot.

The variation from established elevations for slab and floor surfaces should not exceed one-half (1/2) inch in 10 feet, and there should be no offsets or visible waviness in the finished surface.

Steel Trowel Finish When placing concrete that will have a steel trowel finish, the mix should be adjusted to prevent water gain on the surface. The concrete should be brought up evenly to slightly above finished grade and thoroughly compacted by vibrating, tamping, rolling, or other approved means. The top should be struck off to accurately established grade strips or grade blocks, and the surface floated to bring just enough mortar to the surface for troweling.

Fig. 4.18 San Roque three-chute spillway under construction (37 m wide and 450 m long) at 27% slope

The surface should be tested with a straightedge and a template to detect high and low spots which, if found to exceed one-half (1/2) inch in ten (10) feet, should be corrected.

As soon as any moisture sheen has disappeared from the floated surface and the concrete has hardened sufficiently to prevent drawing moisture and fine materials to the surface, the surface should be steel troweled to produce a smooth, hard, uniform finish.

A final steel troweling should be performed after the concrete is hard enough so that no mortar accumulates on the trowel.

Machine finishing producing equally satisfactory results may be used.

4.14 Curing and Protection

Curing is the process to maintain moisture and temperature of fresh concrete during hydration. Curing affects the quality of concrete and its durability.

Curing and protection of concrete should conform to ACI 308R and ACI 301 guidelines. Curing measures should be initiated when the bleed water at the surface evaporates at a faster rate than it can rise to the concrete surface.

Wet curing is recommended and may be achieved by the use of water sprayers, ponding, or covering unformed surface with wet burlap or other approved water-absorbing material. For horizontal and vertical surfaces, moisture-retaining blankets, made of cellulose fabrics that provide constant hydration while maintaining a 100% relative humidity condition, should be considered for effective moist curing.

Liquid membrane-forming curing compound provides an efficient means of curing and for moisture retention. It must conform to ASTM C309 and applied in accordance with the manufacturer's recommendations. If special properties for liquid membrane-forming compounds are required, refer to ASTM C1315.

Curing compounds should be applied on unformed surfaces after completion of finishing and as soon as bleeding has essentially ceased.

For formed surfaces, curing compounds should be applied, immediately, except for surface with defects requiring repairs. For surfaces with defects, the surface should be kept moist to prevent drying.

Curing of Mass Concrete Mass concrete surfaces, where identified on drawings (generally those being 4 feet (1200 mm) in the least dimension), should be moist cured for a minimum of 7 days. Mass concrete should be protected with insulation blankets, of R-value of 1–5 ft*ft^2 F/BTU, determined per ASTM C518. The protection should continue until the concrete surface temperature is within 25° F (14° C) of the average ambient temperature. Thermocouples should be installed in the middle and exterior surfaces of mass concrete to monitor temperatures during curing to verify the differential temperature limits are satisfied.

Curing of Hydraulic Surfaces Concrete surfaces for hydraulic structures should be moist cured for a minimum of 14 days.

Curing of concrete under extreme weather requires controlling the temperature so that hydration continues.

4.15 Removal of Forms

Removal of forms should be in accordance with ACI 301. Minimum strength required prior to removal of forms should meet the requirements of Table 4.6, provided concrete temperatures exceed 50° F (10 °C).

Table 4.6 Minimum strength required for form removal

Member	Minimum strength/minimum time requirements
Nonbearing walls, sides of slabs, beams, girders, and massive foundations	500 psi/24 hours
Columns and bearing walls, with shoring	1000 psi/24 hours
Columns, beams soffit, suspended floor and roof slabs	75 percent of f'_c/14 days

The in-place concrete strength can be estimated in accordance with ASTM C1074, a non-destructive test based on maturity index of a given concrete mixture. The maturity index value considers concrete temperature and curing time. A mix calibration is required. The calibration determines a relationship between maturity and strength for a specific mix. Using a maturity sensor allows the measuring of the in-place temperature of the concrete and then estimates the concrete's strength/maturity through the calibration data previously inputted by the user. This method has been successfully used to facilitate construction schedule.

Case Study: San Roque Multipurpose Project, North Luzon, Philippines

A performance overview of the concrete materials, mix proportions, production, and quality control measures for the dam auxiliary structures, including placing and finishing concrete for the spillway chute, was presented at the ConMat'05 Conference at Vancouver, Canada, August, 2005 [3].

Selection of Concrete Mix Proportions The concrete mixtures were developed for coarse aggregate gradations ranging from 9.5 mm to 75 mm (3/8 inch to 3 inches) maximum size, and slumps range from 25 mm to 200 mm, consistent with the applicable placement requirement and conveying method (bucket/crane, creter crane/conveyors, and pumping methods). Refer to Chap. 2 for concrete mixture designs.

Pumpability Tests On-site field tests were made to verify workability, slump loss, and pumpability of concrete. Concrete was delivered from the central plant in trucks to a yard area where mock-up pumping tests, using straight sections of 100-mm-diameter slickline, were performed. Several iterations requiring adjustments in the sand content, dosages of water reducer, as well as air-entraining agent were necessary for selecting the final mixes suitable for pumping.

Spillway Chute The three-bay spillway, located on the right abutment of the embankment dam, included an approach channel, a gated ogee weir structure, a 450-meter-long spillway chute terminating in a flip bucket, and a downstream channel excavated in rock to return flow to the Agno River. The key features of the spillway included two I-meter-thick and 7-m-high partition walls. Downstream of the ogee, each spillway bay has 37-meter-wide chute (111 m total for three chutes). The slope of the spillway ranges from 1H:3.7V for the first 240-meter length to 1V:21.5H as it enters its terminal structure. The spillway is designed to handle a Probable Maximum Flood (PMF) or a discharge of 12,800 cubic meters (452,000 cubic feet) of water per second. At the end of the chute, a flip bucket was provided.

The spillway chute slab was 0.6 m (2 ft) thick and is provided with a drainage system, between the aerating galleries.

To mitigate cavitation for maximum flow velocities exceeding 40 feet (12 meters) per second, a hydraulic model study was performed that recommended aeration requirement. As a result, each spillway chute was constructed with six aeration galleries extending across the width of the spillway chutes at 55 m interval to provide aeration in the range of 15–17% [4].

A high-performance concrete (HPC) was selected for the spillway concrete with a w/cm ratio of 0.40, including silica fume, to provide erosion resistance for the high-velocity flows. The 5000 psi concrete design mixture included 20% fly ash replacement of cement and 5% silica fume. Due to difficulties in placing and meeting the tight finishing tolerances, the proportions of the HPC were modified with increase in cement from 285 kg/m^3 to 370 kg/m^3 and design strength to 6000 psi while deleting the silica fume. The revised HPC Mixture (AAA2) is shown in Table 2.15 and Fig. 4.18.

Spillway Flatness Tolerances The project's flatness tolerance for the spillway chute was more stringent than the accepted industry tolerance [5] considering that aeration galleries were provided to mitigate cavitation.

Surface irregularity	Specified tolerance	Ref. [5] tolerance (40 ft/sec)
Abrupt irregularity (offset)	0 mm	3 mm
Gradual irregularity (slope)	1.25 mm	6 mm
Gradual irregularity (offset)	6 mm	6 mm

Spillway Construction The spillway's approximately 27 percent slope made concrete placement and equipment and material logistics difficult and were addressed by the following innovations:

1. A specially designed concrete paver, C-700 cylinder finisher, manufactured by GOMACO Corporation, Ida Grove, Iowa, was field tested and modified in the field for compliance with the spillway surface flatness tolerance.
2. Mix S2 (5000 psi) incorporating silica fume was initially selected to enhance the durability of spillway chute. Initial performance of Mix S2 exhibited placing and finishing difficulties; it was replaced with a 6000 psi (56 day) Mix AAA-2, without silica fume. Each chute concrete placement was 37 meters long, by 12 meters wide, 0.6 meters deep.
3. To limit abrupt offsets in the direction of flow, no more than three concrete placements were made for any bay of the chute, between the aeration galleries, resulting in two transverse construction joints in any section.

Concrete was delivered by ready-mix trucks that fed concrete pumps with a telescoping pipe Putzmeister pump, 50 m^3 per hour capacity, system for placing concrete in front of the C700. A special trolley was designed and built on-site to remove the C700 from the completed section, slide it over across the spillway, and line it up into position for the next placement. The concrete placement started at the downstream end and progressed to the upstream higher elevation. Due to the finishing difficulties, the 5000 psi concrete mixture, containing silica fume, was modified eliminating silica fume and increasing the cement content to result in a mixture of

Fig. 4.19 Flip bucket concrete placement at 45-degree slope

6000 psi strength. The actual placement rate was 42 m^3/hr. The chute slab was initially cured with a curing compound and later with water curing to reduce drying shrinkage cracking (Figs. 4.18, 4.19 and 4.20).

4.16 Conclusions and Recommendations

The following conclusions and recommendations are made for mixing, placing, and curing of concrete:

1. For a large project, set up a ready-mixed concrete plant, of adequate capacity, on site.
2. Perform qualification testing of the mixing and transporting equipment, including central mixer, truck mixer, and agitators in accordance with NRMCA recommendations.
3. Establish concrete placement procedures, including the selection of equipment, and rate of placement.
4. For pumping of concrete, establish a correlation between concrete properties at the pump inlet and the pump line discharge, prior to construction.

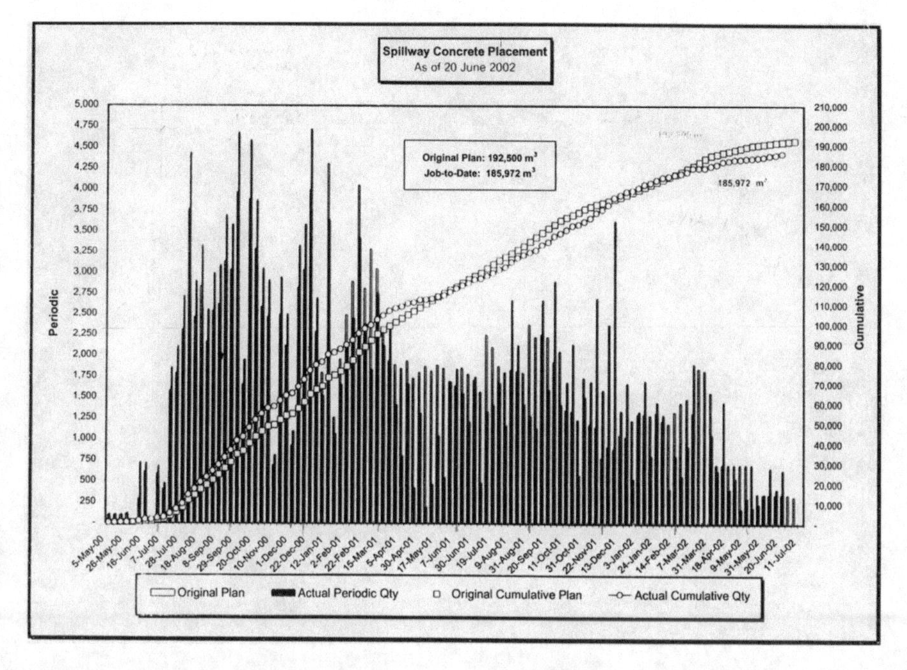

Fig. 4.20 Concrete periodic and cumulative placement (2000–2002)

5. For finishing, attention should be given to concrete mixtures and equipment to meet the specified tolerances.
6. For hot and cold weather, extra attention must be given to assure that concrete is placed without early stiffening or loss of workability.
7. Moist curing is highly recommended to prevent early cracking of concrete. Use moisture-retaining curing blankets, particularly for vertical surfaces, for effectiveness.

References

1. PCA Publication: EB001.16, Chapter 10, Design and Control of Concrete Mixtures
2. Hsieh MC (1982) Slipforming of reinforced concrete shield building. In: Principal Civil Engineer; Ebasco Services, Inc., New York, N.Y. 10048, Jerome R. King, (M.ASCE), Chf. Engr.; Fegles-Power Services Corp., Minneapolis, Minn., ASCE Journal 108,1, 63–73
3. Hasan N (2005) Performance of concrete at san roque dam auxiliary structures. In: Proceedings of Third International Conference on Construction Materials and Mindness Symposium, Vancouver, Canada
4. Engineering Monograph No. 42 (1990) Cavitation on Chutes and Spillway, United States Bureau of Reclamation
5. Concrete manual, US Department of Interior, Bureau of Reclamation, Eighth edition

Chapter 5
Mass Concrete

5.1 Special Requirements for Massive Concrete Structures

Massive Concrete: Structures with large dimensions that require special measures must be taken with temperature and with the generation of heat and attended volume change to minimize cracking. Mass concrete may be defined in terms of the least dimension of the section to be placed. Concrete volume may be considered massive if the least dimension to be placed equals or exceeds 4 feet.

The guidelines for mass concrete mixture should include, but not limited to, the following:

1. Include fly ash (ASTM C618, Type F) up to 25% by weight of to replace cement. The final mix proportions should be established by trial mixes.
2. Type II cement with moderate heat of hydration, or Type IV cement.
3. Largest nominal maximum size coarse aggregate size for concrete, consistent with the reinforcing spacing and concrete cover requirements.
4. Limit the maximum placing temperature at 70 degrees Fahrenheit maximum.
5. The placement schedule for mass concrete shall allow for alternate placing sequence to control volume change and shrinkage cracking. The time interval between adjoining placements shall be specified.
6. The curing of mass concrete shall be performed by extending the wet curing for a minimum of 7 days.

5.2 Historical Mass Concrete Structures

Some of the major mass concrete projects in the USA were concrete dams in the Western USA:

Hoover Dam: constructed in 1935 with over 2.4 million cubic meters of concrete
Grand Coulee Dam: 1942, with over 8.0 million cubic meters of concrete
Shasta Dam, 1945: with over 4.5 million cubic meters of concrete

All the dams used ASTM C150 Type IV low heat cement. Post-cooling in Hoover Dam was the first major application. Concrete was cooled by circulating cold water

N. Hasan, *Durability and Sustainability of Concrete*,
https://doi.org/10.1007/978-3-030-51573-7_5

through the embedded pipes. In addition to post-cooling, the height and sequence of concrete placement was also controlled. All the three dams remained free of cracks and leakage.

In addition to post-cooling of concrete, the exposed surface of the concrete need to be protected by surface insulation to regulate the temperature differential between the concrete surface and the interior. This will reduce thermal stresses and mitigate cracking.

Refer to ACI 207.1R, Table 3.4.2 summarizes the elastic properties of mass concrete for the several major dams constructed in the USA. ACI 207.1R, Table 3.5.1 summarizes the volume change and permeability of mass concrete for several dams constructed in the USA.

This document also provides guidance on placing, consolidation, and curing of mass concrete:

The maximum lift heights for mass concrete ranges from 5 to 7.5 ft for large placements.

Drying shrinkage of mass concrete is low for low-slump lean concrete. Control of temperature rise after placement is an important consideration due to potential risk of cracking and leakage.

Case Study: Thermal Control Plan for Paerdegat Basin Bridge, Brooklyn, NY

A thermal control plan (TCP) for mass placement of concrete footings and superstructure of the new Shore (Belt) Parkway Bridge over Paerdegat Basin Bridge, in Brooklyn, New York, was implemented.

At the request of NYCDOT, URS (a legacy AECOM firm), reviewed the results of thermal modeling and thermal control plan for concrete footings and superstructure members exceeding 1.22 m (4 ft) in thickness and provided recommendations for concrete mix, placement temperature, insulation, cooling, and protection duration requirements for individual concrete members of the new Paerdegat Basin Bridge. The results are shown in Table 5.1.

The thermal control plan (TCP) included the following criteria:

1. Center of concrete temperature was limited to 160F maximum during curing
2. The temperature differential between the center and surfaces of each mass placement was limited to 35F, in accordance with NYC DOT Specifications 555.02.
3. TCP was applicable to sections exceeding 1.22 m (4 ft) thickness, including footings, pier columns, and pier caps. A separate TCP was implemented for each mass placement.
4. All mass placements were provided with insulation, cooling (water) pipes, and sensors for thermal control monitoring. A minimum R2.5 insulation was used for sealing all the exposed surfaces, including protruding reinforcement, during curing.
5. Class HP concrete, supplied by Ferrara Bros., Flushing, with mix proportions and 28-day design compressive strength of 21 MPA (3000 psi) as shown in Table 2.9, Chap. 2.

Table 5.1 Proposed thermal control plan (TCP) for mass concrete recommendations

Structure	Eastbound bridge			Westbound bridge			Recommendations							
								Thermal control plan				Cold weather plan		
	Ref Dwg/ Item #	Thickness m	Thickness ft	Ref Dwg/ Item #	Thickness m	Thickness ft	Concrete mix	Cooling	Tc max	Insulation blanket	Protection duration days	Tc min	Insulation blanket	Protection duration days
							Note 2	Note 3	Note 4	Note 5	Note 6	Note 7	Note 5	Note 8
Abutment	235			258										
Footing (1)	16555.9702	1.5	4.92	16555.9702	1.5	4.92	HP	Yes	75	R2.5	7	45	R2.5	7
Walls (7–9)	16555.9701	1	3.28	16555.9701	1	3.28	HP	No	90	No	7	45	R2.5	7
Walls (13–15)	16555.9701	0.48–0.60	2	16555.9701	0.48	2	HP	No	90	No	3	45	R2.5	7
Pier 1	241			263										
Footing (1)	16555.9702	2.25	7.4	16555.9702	2.45	8.5	HP	Yes	75	R2.5	7	45	R2.5	7
Columns (2–4)	16555.9701	1.5–2.16	7	16555.9701	1.75–2.75	5.75–9.0	MP	Yes	75	R2.5	7	45	R2.5	7
Pier cap (5)	16555.9701	2	6.5	16555.9701	2.5	8.2	MP	No	75	R2.5	14	45	R2.5	7
Pier 2	244			266										
Footing (1)	16555.9702	2.6	8.5	16555.9702	2.45	8.5	HP	Yes	75	R2.5	7	45	R2.5	7
Columns (2–4)	16555.9701	1.75–2.615	7	16555.9701	1.75–2.75	5.75–9.0	MP	Yes	75	R2.5	7	45	R2.5	7
Pier cap (5)	16555.9701	2.5	8.2	16555.9701	2.5	8.2	MP	No	75	R2.5	14	45	R2.5	7
Pier 3	247													
Footing (1)	16555.9702	2.6	8.5				HP	Yes	75	R2.5	7	45	R2.5	7
Columns (2–4)	16555.9701	1.75–2.615	7				MP	No	75	R2.5	7	45	R2.5	7
Pier cap (5)	16555.9701	2.5	8.2				MP	No	75	R2.5	14	45	R2.5	7
Pier 4	250													
Footing (1)	16555.9702	2.25	7.4				HP	Yes	75	R2.5	7	45	R2.5	7
Columns (2–4)	16555.9701	1.5–2.16	7				MP	No	75	R2.5	7	45	R2.5	7
Pier cap (5)	16555.9701	2	6.5				MP	No	75	R2.5	14	45	R2.5	7

5.3 Field Setup

Based on thermal modeling data, the following features were incorporated for cooling and monitoring of mass concrete placements:

1. Cooling Pipes: 3/4" PEX tubing (spacing: 1 m c/c) tied to reinforcing steel (Fig. 5.1).
2. The cooling pipes were connected to a 4-inch manifold for water inlet and outlet lines (Figs. 5.2 and 5.3).
3. The cooling pipes were filled with water before concreting (Tables 5.2 and 5.3).
4. A total of ten temperatures loggers (five primary and five secondary) were installed in each mass concrete placement as shown below: water inlet and outlet logger (four), ambient air logger (two), surface logger (two), and center Logger (two).
5. Cold water was circulated at the rate of 5–6 gpm (Fig. 5.4). The inlet and outlet water temperatures ranged from 46 °F to 86 °F, respectively (Table 5.4). Temperatures were continuously monitored hourly and downloaded daily. The cooling was continued until the temperature differential (core to surface) was less than 35 °F.
6. Upon termination of the TCP, the cooling pipes were filled with non-shrink grout (Figs. 5.3 and 5.4).

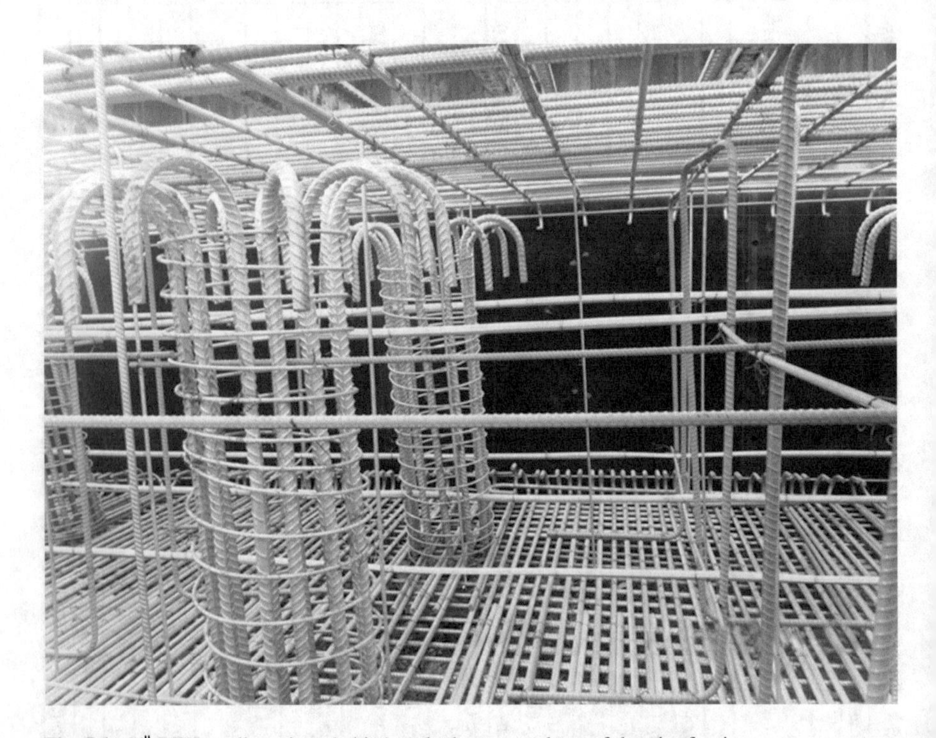

Fig. 5.1 3/4 PEX cooling pipes (white) at the bottom and top of the pier footing

Fig. 5.2 PEX cooling pipes outside the forms

Fig. 5.3 PEX pipe connection to water manifold

Table 5.2 Cooling water summary for the mass concrete pier footings

	Date placed	Date terminated	Thickness m (ft)	Cooling water temp (F) range
East bound				
Pier 1 footing	12/11/2010	12/16/2010	2.25 (7.38)	52–73
Pier 2 footing	1/10/2011	1/18/2011	2.60 (8.50)	34–61
Pier 3 footing	2/16/2011	2/24/2011	2.60 (8.50)	46–79
Pier 4 footing	12/22/2010	1/3/2011	2.25 (7.38)	48–68
West bound				
Pier 1 footing	12/8/2011	12/15/2011	2.6 (8.50)	52–73
Pier 2 footing	5/12/2012	5/21/2012	2.6 (8.50)	55–86

For the piers, the maximum in-place temperature was 147 °F and reached 27–40 hours after the placement. There was a good correlation between the predicted and actual concrete temperatures. The maximum differential temperature ranged from 25 °F to 29 °F. The difference between the predicted and actual concrete temperatures was less than 10 °F, as shown in Figs. 5.5, 5.6, 5.7, 5.8, and 5.9.

Results The TCP for Paerdegat Basin Bridge concrete foundations and superstructure was implemented successfully during the 2011–2012 construction period. In accordance with the specified design criteria, the results confirmed compliance with the maximum and differential concrete temperatures limits. The maximum concrete temperature recorded for any mass concrete placements was 147 °F, which was well below the specified 160 °F limit. The removal of concrete protection and termination of cooling time for concrete footings, which was generally dictated by the low ambient air temperatures, occurred after 8–9 days and were in accordance with the predicted TCP data.

The TCP maximum concrete and differential temperatures, with cooling, provided and exhibited a close relationship with the predicted TPC temperatures for the individual placements, as shown in Table 5.4 and Fig. 5.5.

Case Study: Sidney A. Murray Jr. Hydroelectric Station (1985–1990) (Ref. [1])

The Sidney A. Murray Jr. Hydroelectric Station, a low-head run of river powerplant, is located about 40 miles south of Vidalia, LA, on the Mississippi River. The powerhouse itself is the world's largest prefabricated powerplant structure (PPS). Upon arrival at the site, the PPS was lowered and set onto reinforced concrete foundation mat. After site dewatering and cleaning the cavities, over 110,000 cu. yd of concrete was placed (Fig. 5.10).

The design required a composite steel/concrete performance of the structure, requiring concrete to fill each compartment completely without voids. To minimize

Table 5.3 Thermal control plan for Class HP concrete – predicted vs actual differential temperatures

			With cooling pipes – predicted				With cooling pipes – actual						
			Initial	Initial Air	Predicted	Predicted	Conc	Air	Water	Temp, F	Temp, F	Delta	
Location	Thickness	Insulation	Concrete	Ambient	Conc	Diff		Ambient		Interior	Exterior	Diff	
	Meters		Temp, F	Temp, F	Temp, F	Temp, F	Temp, F	Temp, F	Temp, F	Temp, F		Temp, F	Time
EB piers	2.25	R2.5	50	50	121	18	68	59	46	68	64	4	1 hour
Pier 1	7.4 ft		60	50	132	20		51	62	147	127	30	27 hours
Footing			70	50	**144**	**23**		23	53	133	108	25	72 hours
								23	50	127	100	27	85 hours
EB piers	2.6	R2.5	50	50	119	18	68	34	68	68	63	5	0
Pier 2	8.5 ft		60	50	129	20		26	88	133	114.8	24	40 hours
Footing			70	50	**141**	**23**		30	72	122	97	25	72 hours
								32	63	102	82	20	120 hours
EB	2.1	R2.5	50	50	**121**	**20**	50	28	37	50	50	0	0
W abutment	7 ft		60	50	132	23		36	39	117	102	15	30 hours
Stem wall			70	50	144	26		41	43	117	95	22	48 hours
								34	42	108	88	20	68 hours
EB	2.1	R2.5	50	50	121	20	64			64	64	0	0
West abut			60	50	**132**	**23**				129	99	30	30 hours
Center wall			70	50	144	26				125	91	34	48 hours
										113	82	31	72 hours
										100	68	32	96 hours
										89	62	27	120 hours
WB	1.5 × 1.5	R2.5	50	50	89	9	59	57	48	59	59		0
Pier 1	Aver		60	50	**98**	**12**		27		106	90	16	24 hours
S1 column			70	50	109	14		27		113	101	12	48 hours
								27	90	75	15	72 hours	

(continued)

Table 5.3 (continued)

			With cooling pipes – predicted				With cooling pipes – actual						
			Initial	Initial Air	Predicted	Predicted	Conc	Air	Water	Temp, F	Temp, F	Delta	
								27	86	75	11	89 hours	
EB	1.75 × 1.75	R2.5	50	50	89	9	73	46	43	73	70	3	0
Pier 2	Aver		60	50	98	12		57		144	129	25	24 hours
S2 Column			70	50	**109**	**14**		72		129	109	20	48 hours
								50		127	113	14	72 hours
								46		115	104		96 hours

Fig. 5.4 Valves and flow gauges at the manifold

Table 5.4 Results of maximum predicted and actual temperatures and variances

Structure	Maximum temperature rise (F)			Maximum differential temperatures (F)		
	Predicted	Actual	Variance	Predicted	Actual	Variance
Pier footings	144	147	+3	23	29	+6
Column S2	109	129	+20	14	20	+6
Column S1	109	113	+4	14	12	−2

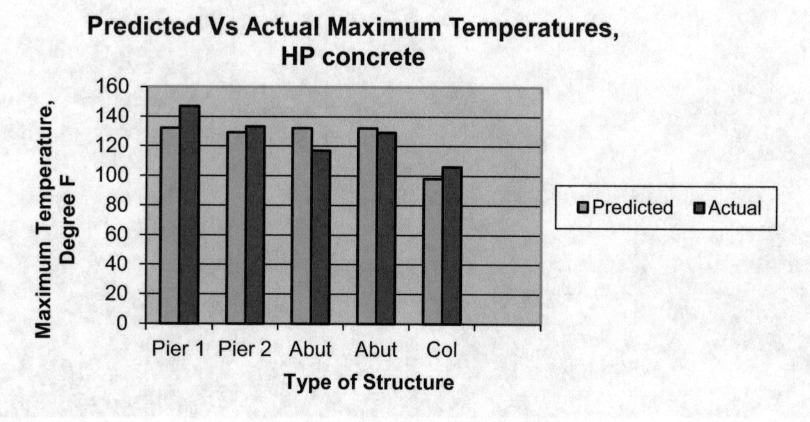

Fig. 5.5 Predicted vs actual maximum temperatures for HP concrete

Fig. 5.6 Eastbound bridge lanes completed

Fig. 5.7 Installation of girders at westbound bridge section

Fig. 5.8 Demolishing of westbound bridge section

Fig. 5.9 Completed eastbound (foreground) and westbound Paerdegat Basin Bridge

Fig. 5.10 PPS being fabricated at New Orleans shipyard

thermal cracking in concrete from heat of hydration a maximum temperature of 118 °F on concrete during setting and hardening was specified. The design compressive strength of concrete at 28 or 90 days for concrete containing fly ash was established at 3000 psi.

The selected concrete mixture proportions for the superplasticized concrete (SCC) are shown in Table 2.16, Chap. 2.

Tests were performed on concrete at the batch plant to verify compliance with the specification requirements. The SCC was tested for temperature, flowability, and cylinders for strength tests every 150 yd^3 delivered. Due to good quality control, deviations from the specification were kept to a minimum.

5.4 Concrete Placement Procedure

Tests were performed to establish the maximum unconfined concrete drop that could be sustained without segregation of the SCC. Height of vertical unconfined drop varied from 5 ft to 15 ft. The results indicated that there was no segregation of the mixture when concrete was dropped from 15 ft height. The placement rate was limited to 6 feet per hour. It was closely monitored through instrumentation on steel plating to assure the lateral pressures were compatible with the design.

Based on a study performed to predict both the amount and age of peak temperature rise associated with heat of hydration for various volume to surface ratios.

Table 5.5 Placement by concrete mixture type

Concrete mixture	Quantities cu. yd
3A.2S	44,131
3A.2	11,716
3A.8S	47,354

Placing temperatures, mixture, and a maximum 68 °F for fresh concrete placing temperature were specified for all placements. A minimum of 48 hours was imposed for selective massive placements.

The test results also indicated that for superplasticized concrete, the combination of drop height and horizontal flow must be limited to 20 feet maximum to prevent segregation.

The concrete was batched from on site, 100 yd ^{3}/hour plant, was equipped, for cooling concrete, with nitrogen injection system, and delivered to the site in Transit Maxon dump trucks. Concrete was placed by pumping and by gravity through hopper-drop chute, in a continuous 24-hour operation from Monday to Saturday in three shifts.

About one-half of the concrete was placed from top of the powerhouse and dropped it into hoppers via 8-inch diameter flexible vinyl drop chutes. The drop chutes were cut and removed as placement progressed. The rest of the concrete was pumped through a Schwing BPL 900 GDR pump (capacity 107 yd^3/hr with a 5-inch diameter slickline to the lower powerhouse). A standby pumping equipment (Western American W45) was available.

Table 5.5 presents a summary of concrete placed in the PPS over a period of 5 months.

Case Study: Rainbow Redevelopment Project, Great Falls, MT

Powerhouse Mass Concrete

The mass concrete mixture proportions and strength data are shown in Tables 5.6 and 5.7.

For the powerhouse mass concrete placements (Fig. 5.11), a thermal control plan (TCP) was developed by the contractor (Walsh in consultation with CTL Group, Skokie, IL), based on the following acceptance criteria:

1. Maximum temperature limit: 160 °F
2. Maximum differential temperature in a lift: 35 °F

Plastic cooling pipes were installed at the bottom of each lift, and temperature sensors (Intellirock) were placed in the middle, side, and top surfaces of each lift. After placing concrete, the surfaces of concrete, and the protruding steel to a distance of 3 feet from the surface, were protected with R-2.5 insulating blankets, and

Table 5.6 4000 psi mass concrete mixture proportions (per cubic yard unless noted)

Designation	Mix 4000 M
Water/cementitious material ratio	0.49
Cement Type II/V, lb	388
Fly ash Class F, lb	129
Water, lb	253
Sand, SSD, lb	1249
#467 stone, SSD, lb	1866
MB micro air, oz	4.1
MB Polyheed 997, oz	25.9
MB Rheobuild 1000, oz	
Ambient temperature, F	
Concrete temperature, F	69
Slump, inches	3.00
Air content, %	4.8
Unit Weight, pcf	143.6
Cylinder compressive strength, psi	
3 days	
7 days	3260
28 days	4310

Table 5.7 4000 psi Mass Concrete Mixture Proportions Strength Data (psi)

Designation	14-day value	28 day value
Compressive strength (15 days)	3023 psi	3820 psi
Splitting tensile strength	367 psi	
Coefficient of thermal expansion	5.45*10^−6/F	

cooled with water circulating through the plastic pipes for a minimum duration of 3 days, or until the TCP acceptance criteria was met.

Concrete quantities for the powerhouse lifts 2 thru 8, lift heights, and placement rates utilizing two Putzmeister pump trucks (target capacity per pump: 92 cy/hour) are shown in Table 5.8. The acceptance criteria

Table 5.9 shows the predicted and actual temperatures for selected mass concrete lifts in the powerhouse. A peak temperature of 129 °F was reached in lift 2 interior after 61.25 hours. The maximum differential temperature reached in any lift was 34 °F and was within the TCP acceptance criteria (Figs. 5.12, 5.13, and 5.14).

Compressive strengths for the mass concrete mix 5.5 are presented in Figs. 5.15 and 5.16.

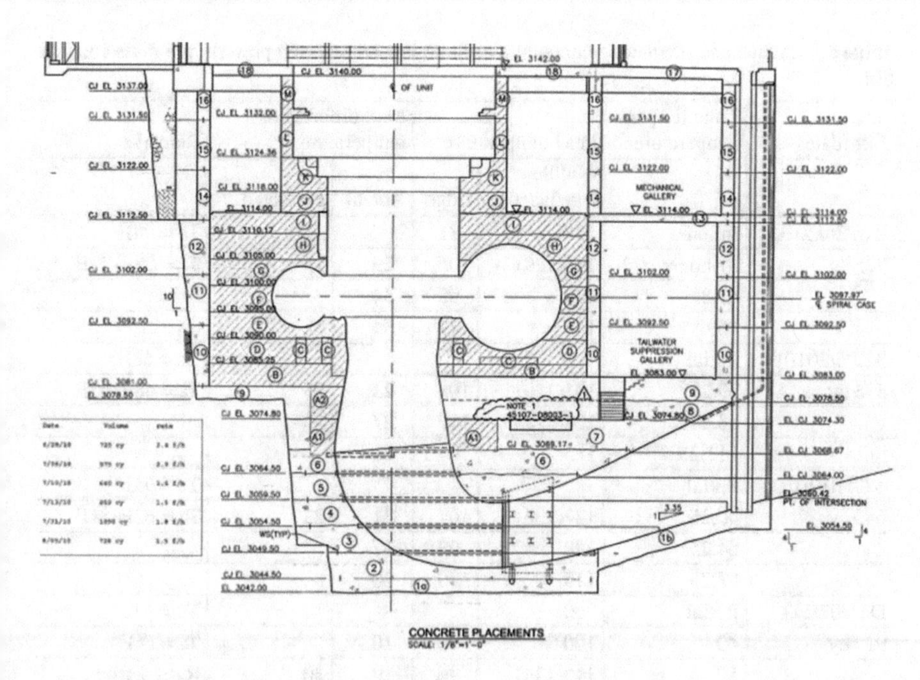

Fig. 5.11 Rainbow Powerhouse mass concrete lifts 2–6 around the draft tube

Table 5.8 Powerhouse mass concrete lifts and placement rates

Lift (date)	Quantity(actual) (yd^3)	Lift height (ft)	Placement rate (max) Yd^3/hour	Placement rate (actual) (ft/hr)
2 (6/28/2010)	665(725)	5.00	133	2.4
3 (7/06/2012)	511(575)	5.00	102	2.0
4 (7/10/2010)	604(640)	5.00	121	1.5
5 (7/17/2010)	859(864)	5.00	172	1.5
6 (7/23/2010)	916 (851)	4.67	196	
7 (7/31/2010)	1286(1090)	5.63	228	1.8
8 (8/10/2010)	728(800)	3.70	197	1.5
Subtotal	5569 (5545)			

Table 5.9 Actual and predicted temperature differentials for selected powerhouse mass concrete lifts

Lift (date)	Time for peak temperature	Peak temperature		Max differential temperature		Remarks
		Middle (predicted)	Edge	Actual	Predicted	
2 (6/28/2010)	Initial	77	81			Tc = 74F
725 cy	37 hours	135(129)	106	29	25	Ta = 53F–72F
	61.25	129	95	34		
	109	113	79	34		
3 (7/6/2010)	Initial					Tc = 68F
575 cy	39.25	131(129)	108	23	25	Ta = 48F–63F
	63.25	127	100	27		
	111.25	115	94	21		
4 (7/10/2010)	Initial					Tc = 75F
640 cy	40.25	127(136)	104	31	23	Ta = 62F–81F
	64.25	120	93	27		
	88.25	115	90	20		
D (5/07/11)	Initial	60				Tc = 61F
920 cy	40	100	90	10		Ta = 66F
	50	115(124)	96	19	30	Rate: 1 ft/hr
	72	122	96	26		(0.75 ft/hr (actual)
	144	104	100	4		
E(05/11/11)						Tc = 63F
510 cy	Initial	70	66			Ta = 49F
	50	115(117)	105	10	31	Rate: 1 ft/hr
	72	120	105	15		(0.5 ft/hr (actual)

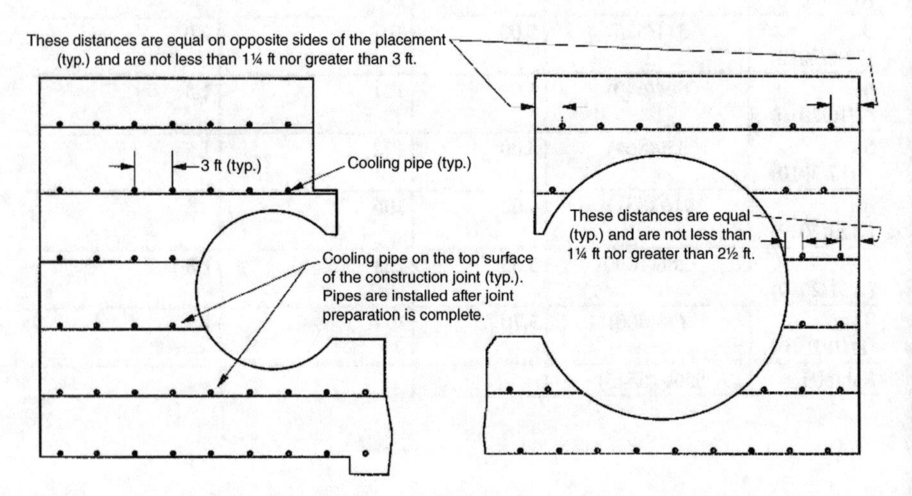

Fig. 5.12 Location of cooling pipe in each 5-ft lift

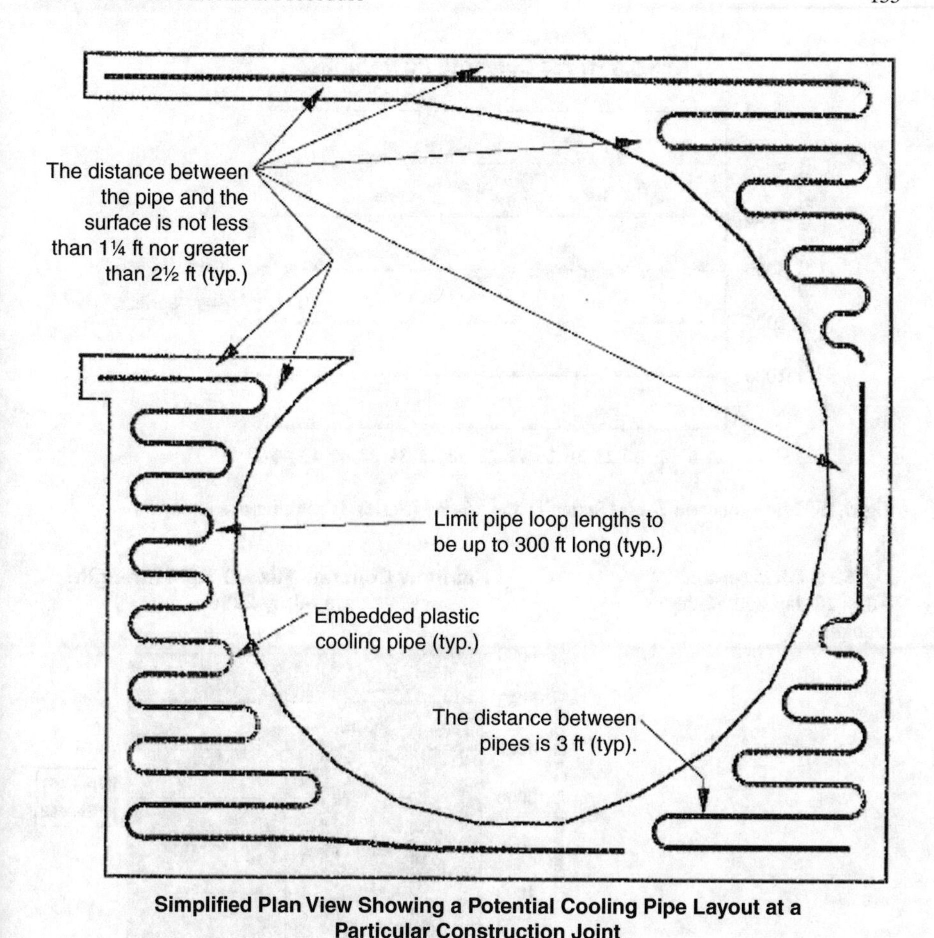

Fig. 5.13 Layout of plastic tubing around scroll case for cooling mass concrete after placement

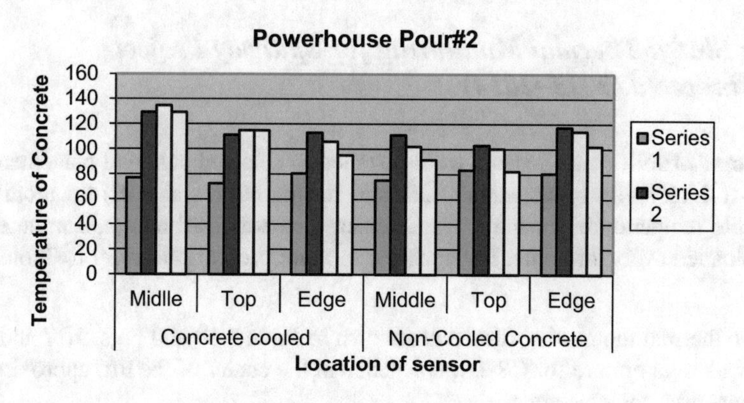

Fig. 5.14 In-place concrete temperature of lift 2 mass concrete after placement. (Series1:0 hrs, Series 2:13 hrs, Series3:a 37 hrs, Series 4: 61 hrs)

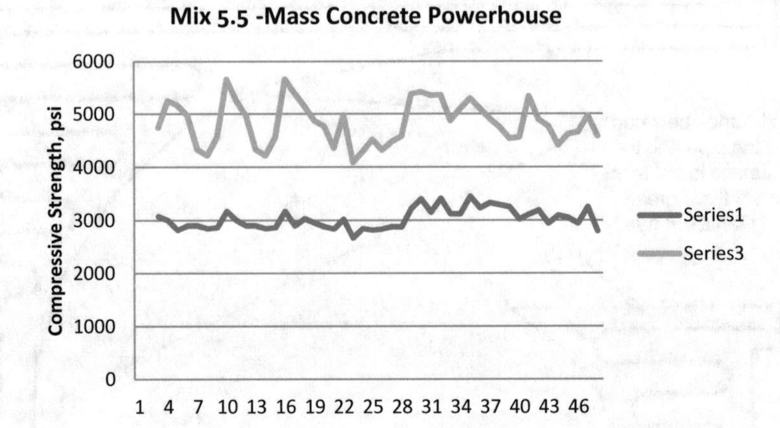

Fig. 5.15 Mass concrete 7-day (Series 1) and 56-day (Series 3) compressive strengths

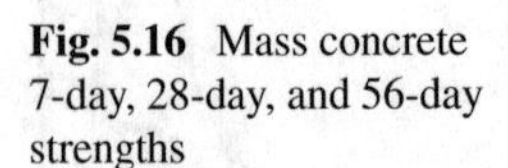

Fig. 5.16 Mass concrete 7-day, 28-day, and 56-day strengths

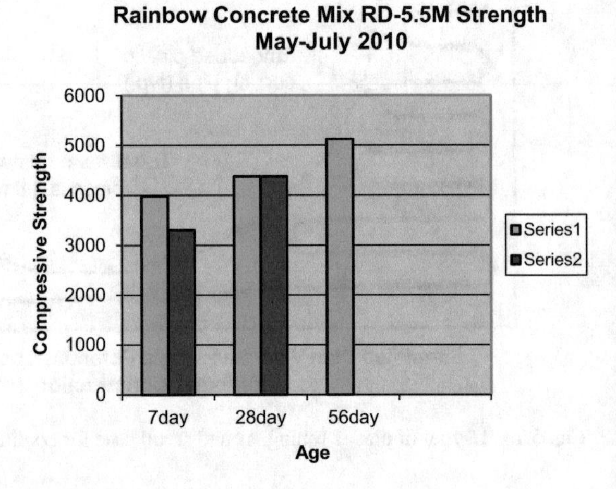

Case Study: Thermal Monitoring for Spillway Project in Minnesota (2013–2014)

Thermal Monitoring Sensors were installed in selected concrete placements for the 4-ft thick spillway chute slab (CS) and stilling basin slab (SB) for monitoring concrete temperatures during curing. Sensor data was read every 30 minutes and downloaded every 24 hours. For concrete mixture proportions, refer to Table 2.22, Chap. 2.

The thermal monitoring results are shown in Table 5.10 and Figs. 5.17 and 5.18. The peak temperature for CS-1 was reached at the center of the lift, approximately 69 hours after the placement.

Table 5.10 Thermal monitoring of concrete

Location	Placement	Initial temperature ° F		Maximum temperature
		Concrete	Air	
21 feet east of CL	SB-1	39.2	14	115.7
At center	CS-1	58.1	14	112.1 (69 hours)
At edge	CS-1	60.8	14	94.1 (50 hours)

Note: sensors were located 18 inches below the surface

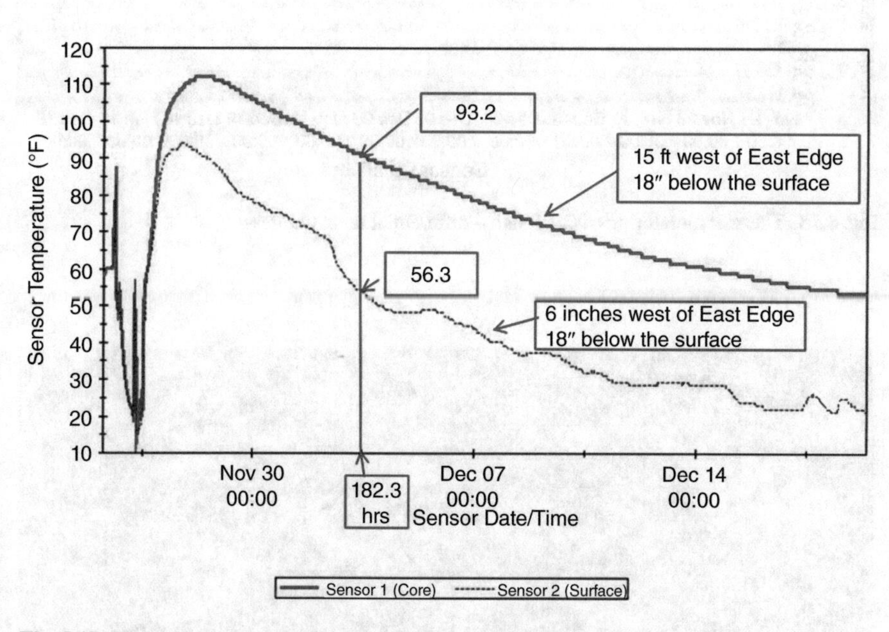

Fig. 5.17 Thermal monitoring of chute slab CS-1 at the center and edge

Case Study: Chickamauga Lock Replacement on Tennessee River, Chattanooga, TN

Chickamauga Lock is located on Tennessee River at Chattanooga, TN. Laboratory testing of concrete for the Chickamauga Lock replacement were performed in accordance with ACI 211.1 in 2018. Several trial mix proportions for mass concrete, structural concrete, second-stage concrete, and drilled shaft concrete were tested. Concrete mixture proportions were selected to attain the required slump, air entrainment, unit weight, and compressive strength properties of the specification (Fig. 5.19).

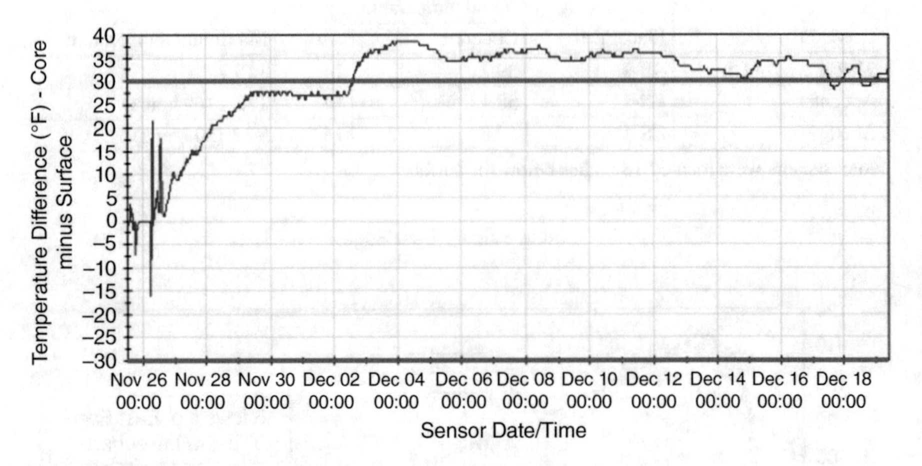

Fig. 5.18 Thermal monitoring of CS-1 slab – differential temperatures

Fig. 5.19 Chickamauga Lock under construction (2020). (photo courtesy of Aerial Innovations)

Cement for Mass Concrete The cement, manufactured at the LafargeHolcim plant, conforms to an ASTM C150 Type II (MH), with low alkalis and zero C_3A content. It mitigated the ASR expansion phenomenon associated with the potentially reactive aggregates (Refer to Chap. 2). The cement provided a high resistance to the sulfates, present in the local soils.

A typical Type II (MH) cement mill test report results are presented in Table 5.11. This cement is being used for Chickamauga Lock at Chattanooga, TN. Typical Type I/II Buzzi Unicem cement mill test results are shown in Table 5.12.

Table 5.11 Chickamauga LafargeHolcim Type II (MH) cement mill test results

Requirements	ASTM C150 Type II (MH)	CMTR report Feb 2017
Chemical requirements		
SiO_2, %	20.0 min.	21.7
Al $_22O_3$, %	6. 0 max	2.9
Fe_2O, %	6. 0 max	4.5
CaO, %		64.1
MgO, %	6. 0 max	3.1
SO $_3$, %	3. 0 max	2.9
LOI, %	3. 0 max	1.0
Insoluble residue, %	1.50 max	0.19
Na O equivalent, %	0. 60 max	0.25
C_3S, %		62
C_2S, %		16
C_3A, %	8. 0 max	0
C $_4$AF, %		14
C $_3$S + 4.75*C $_3$A, %		62
Physical requirements		
Blaine fineness, m2/kg (air permeability)	260 min	303
Minus 325 mesh		82.5
Vicat set time		
Initial, minutes	45 min	89
Final, minutes	375 min	325
False set, %	50 min	72
Heat of hydration, cal./g	70	63
Air content, %	12 max	6.0
Autoclave expansion, %	0.80 max	−0.02
Mortar expansion, % (C11038)	0.02 max	0.000
Compressive strength, psi		
1 day		830
3 day	1450 psi	2000
7 day	2470 psi	3070
28 day		4870

Table 5.12 Chattanooga Buzzi Unicem Type I/II cement mill test results

Requirements	ASTM C150 Type I/ II	CMTR report Sep 2017
Chemical requirements		
SiO_2, %	20.0 min.	19.8
Al_2O_3, %	6. 0 max	4.4
Fe_2O, %	6. 0 max	3.2
CaO, %		62.5
MgO, %	6. 0 max	3.2
SO_3, %	3. 0 max	2.8
LOI, %	3. 0 max	1.0
Insoluble residue, %	1.50 max	0.39
Na_2O equivalent, %	0. 60 max	0.58
C_3S, %		61.6
C_2S, %		10.4
C_3A, %	8. 0 max	6.2
C_4AF, %		9.8
$C_3S + 4.75*C_3A$, %		90.9
Physical requirements		
Blaine fineness, m^2/kg (air permeability)	260 min	389
Minus 325 mesh		95.5
Vicat set time		
Initial, minutes	45 min	89
Final, minutes	375 min	325
False set, %	50 min	61
Heat of hydration, cal/g		
Air content, %	12 max	7.2
Autoclave expansion, %	0.80 max	0.03
Compressive strength, psi		
1 day		2343
3 day	1450 psi	3912
7 day	2470 psi	4952
28 day		6507

5.5 Recommendations

For successful mass concrete placements, a thermal modeling and control plan should be developed prior to construction. The thermal control should be focused on the following criteria:

1. The concrete mixture proportions should be selected with Type II cement, with 25% fly ash and/or GGBFS replacement, and with maximum nominal size of coarse aggregate, consistent with thickness and method of concrete placement. The design strength should be 3000 psi at 28 days or preferably based on 56 days.

2. Limit the placement temperature to 70 °F maximum.
3. The maximum in-place temperature should be 158 °F.
4. Install temperature sensors in each lift to monitor concrete temperatures.
5. Protect the exposed surfaces of concrete and reinforcing steel, with insulation. The insulation R-value should be based on the thermal modeling.
6. If required, cooling pipes may be installed for post cooling of concrete.
7. The protection and cooling of concrete should be continued until the temperature differential between the interior and the surface is less than 35 °F. Time interval between lifts should be 72 hours minimum, if temperature sensors are not installed.
8. Upon termination, the cooling pipes should be filled with non-shrink grout.

It is recommended that selection of mass concrete mixture proportions should give due consideration to compressive strength and loading requirements, and preparation of concrete placement procedures to limit the temperature rise and temperature differences between the center and the surface, and insulation requirements. For planning purposes, it is recommended to allow adequate time (preferably 3–6 months) for this effort, consistent with the specifications, site conditions, and schedule requirements.

It is recommended that the design compressive strength requirement for mass concrete be relaxed from 28 to 56 days age. This will reduce the cementitious content of the mixture and in turn should reduce the duration of thermal control by approximately one-third than that required.

References

1. Hasan N (1991) Superplasticized concrete for a prefabricated powerplant structure. In: Waterpower '91, Proceedings of the international conference on hydropower, vol 3, Denver, CO, pp 1529–1538

Chapter 6
Self-Consolidating Concrete (SCC)

6.1 General

ACI 237-07, "Report on Self-Consolidating Concrete," defines self-consolidating concrete (SCC) as a highly flowable, nonsegregating concrete that can spread into place, fill the formwork, and encapsulate the reinforcement without any mechanical consolidation. In general, SCC is concrete made with conventional concrete materials and may include a viscosity modifying admixture (VMA). Its economic benefit lies in eliminating labor and equipment costs for consolidation and finishing, while providing enhanced performance during service life including compressive strength and durability.

SCC mixtures can exhibit excellent resistance, with favorable air-void systems and spacing factor less than 0.008 in. (0.2 mm), to freezing and thawing and to deicing salts.

The use of SCC has gained acceptance since the late 1980s in Japan, where initially it was developed to ensure proper consolidation where durability and service life were of concern. The use SCC has grown dramatically in the USA. The SCC mixtures typically incorporate coarse aggregates (3/4-inch NMS or smaller), water reducing admixtures (WRA), and/or high range water reducers (HRWR), conforming to ASTM C494.The dosage rates of WRA and HRWR are established by trial concrete batches, tested in accordance with AC 211.1 and verified from the on-site concrete plant.

The flowability characteristics are determined by a slump flow test, based on ASTM C143 with modifications, as described in ASTM C1611. The typical slump flow values range from 20 to 30 inches. When tested by ASTM C1611, the stability of the SCC can be observed visually for a qualitative assessment as VSI values, shown in Figs. 6.1, 6.2, 6.3, and 6.4 (Courtesy of BASF).

N. Hasan, *Durability and Sustainability of Concrete*,
https://doi.org/10.1007/978-3-030-51573-7_6

Fig. 6.1 Highly stable. (Photo sources: Courtesy of BASF)

Fig. 6.2 Unstable – slight mortar halo. (Photo sources: Courtesy of BASF)

Fig. 6.3 Stable. (Photo sources: Courtesy of BASF)

Fig. 6.4 Highly unstable – center of mass segregation (Photo sources: Courtesy of BASF)

Visual Stability Index (VSI) value	Criteria
Highly stable (Fig. 6.1)	No evidence of segregation
Stable (Fig. 6.2)	No evidence of segregation and slight bleeding sheen
Unstable (Fig. 6.3)	A slight mortar halo
Highly unstable (Fig. 6.4)	Segregation – large mortar halo >0.5 in and large aggregate pile in the center of the mass

Due to its highly flowable nonsegregating characteristics that fill the formwork without mechanical consolidation, SCC main economic benefit lies in eliminating labor and equipment costs for consolidation, while providing enhanced performance during service life including compressive strength and durability (Figs. 6.5 and 6.6).

SCC has a drying shrinkage of less than 0.042% at 28 days, when tested in accordance with ASTM C157. The SCC mixes incorporate water-reducing admixtures (WRA) and/or high range water reducers (HRWR), conforming to ASTM C494.

6.2 Self-Consolidating Concrete Mixtures

The concrete mix proportions should be established to meet the design requirement of ACI 237R, ACI 301 ACI 318, and ACI 211.1.

The SCC may be placed by hopper or buckets, or by pumping, depending upon the site conditions and rate of concrete placement.

Fig. 6.5 #8 Coarse aggregate for SCC

Fig. 6.6 SCC slump flow test ASTM C1611

Hardened Properties Refer to ACI 237R-07 for details on design of SCC mixtures, physical properties, and placement guidelines.

Typical concrete mixture proportions for SCC concrete with compressive strength ranging from 4000 to 6000 psi minimum at 28 days or 56 days (with fly ash). Typical SCC trial mixture proportions are listed in Table 6.1.

SCC is being used for a variety of concrete structures due to ease of placement in heavily reinforced and congested areas without consolidation. SCC, being more flowable, eliminates large air voids, which may be present in conventional concrete, and it improves mechanical properties and reduces permeability of concrete. This assures long-term durability. Due to its flowability, however, the formwork must be leak-tight and designed to resist the lateral hydrostatic pressures, and/or the rate of concrete placement needs to be monitored.

Case Study: St. Lucie Nuclear Plant Intake Velocity Cap

In 1992–1993, underwater repairs to, and rehabilitation of, existing reinforced concrete velocity caps of the circulating water intake structure at St. Lucie Powerplant, Fort Pierce, Florida, were made utilizing high-performance SCC in a marine environment. Use of this repair technique avoided the necessity of constructing a cofferdam for repair work in the dry, and thus minimized interruption to plan operation, and resulted in considerable savings.

Table 6.1 Typical SCC proportioning trial mixture summary[1]

Absolute volume of coarse aggregate	28–32%
Coarse aggregate size	3/8 inch, 3/4 inch
Paste fraction	34–40%
Typical w/cm ratio	0.32–0.45
Typical cement content (powder content)	650–800 lb./cy (386–475 kg/m^3)

[1]ACI 237R, Table 4.2

Mix proportions for the high-performance concrete included cement, fly ash, silica fume, and anti-washout (also known as viscosity modifying) admixture as well as high-range water-reducing and set-retarding admixtures and were tested extensively to achieve a flowable and self-consolidating concrete while assuring early compressive strength requirement. Large-scale mock-up tests, utilizing both tremie and pumping methods, were conducted to simulate underwater placement in the surf zone and to select the actual concrete placing method and rate of placement and to identify surface preparation and protection requirements.

Construction procedures for the new reinforced concrete slabs involving approximately 3000 yd^3 precast and tremie concrete utilizing a barge-mounted concrete batch plant, quality control, and post-placement inspection measures are also discussed in Ref. [1].

The SCC mixture proportions, and design strength requirements, are presented in Table 6.2 and Fig. 6.7. The early age design compressive strength was 1000 psi minimum at 24 hours.

The project demonstrated that SCC can be successfully placed underwater by tremie pipes with production rates of 72 cy/hr. The use of an AWA and silica fume allowed placement in ocean surf zone with a minimal loss of material and segregation. The use of HPC technology along with the innovative design solution greatly simplified the reconstruction of St. Lucie Nuclear Plant intake velocity caps (Table 6.3).

Table 6.2 High-performance SCC for surf zone with AWA-6000 psi at 28 days

Designation	Quantities
Water/cementitious material ratio	0.40
Cement Type I-II, lb./cy (kg/m^3)	600 (357)
Fly ash Class F, lb./cy (kg/m^3)	90 (53.55)
Silica fume, lb./cy (kg/m^3)	43 (25.6)
Water, lb./cy (kg/m^3)	293 (174.5)
Sand, SSD, lb./cy (kg/m^3)	1467 (872.9)
#8 stone, SSD, lb./cy (kg/m^3)	1550 (922.2)
AEA, oz./ yd. (ml/m^3)	5.5 (64)
HWRA, gal/yd. (ml/m^3)	1.9 (260)
AWA, gal/yd. (ml/m^3)	1.6 (65)
Initial tests after batching (at plant)	
Concrete temperature, ^{0}F	55+/−5
Initial slump, inches minimum	9
Air content, %	
Unit weight, lb./cf.	145
Initial set time	9 hours minimum
Minimum compressive strength	
24 hours	1000 psi
7 days	4000 psi
28 days	6000 psi

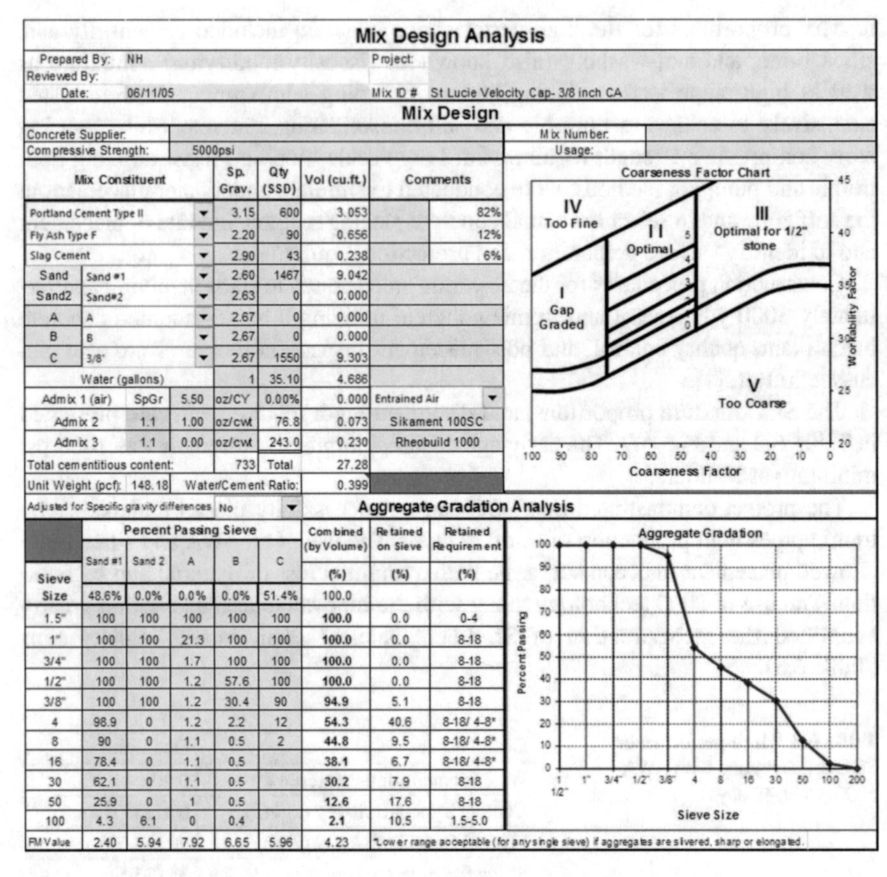

Mix Design Analysis

Prepared By:	NH	Project:	
Reviewed By:			
Date:	06/11/05	Mix ID #	St Lucie Velocity Cap- 3/8 inch CA

Mix Design

Concrete Supplier:		Mix Number:	
Compressive Strength:	5000psi	Usage:	

Mix Constituent		Sp. Grav.	Qty (SSD)	Vol (cu.ft.)	Comments
Portland Cement Type II		3.15	600	3.053	82%
Fly Ash Type F		2.20	90	0.656	12%
Slag Cement		2.90	43	0.238	6%
Sand	Sand #1	2.60	1467	9.042	
Sand2	Sand#2	2.63	0	0.000	
A	A	2.67	0	0.000	
B	B	2.67	0	0.000	
C	3/8"	2.67	1550	9.303	
Water (gallons)		1	35.10	4.686	
Admix 1 (air)	SpGr 5.50 oz/CY		0.00%	0.000	Entrained Air
Admix 2	1.1 1.00 oz/cwt		76.8	0.073	Sikament 100SC
Admix 3	1.1 0.00 oz/cwt		243.0	0.230	Rheobuild 1000
Total cementitious content:		733	Total	27.28	
Unit Weight (pcf): 148.18	Water/Cement Ratio:			0.399	

Adjusted for Specific gravity differences: No

Aggregate Gradation Analysis

Sieve Size	Sand #1	Sand 2	A	B	C	Combined (by Volume) (%)	Retained on Sieve (%)	Retained Requirement (%)
	48.6%	0.0%	0.0%	0.0%	51.4%	100.0		
1.5"	100	100	100	100	100	100.0	0.0	0-4
1"	100	100	21.3	100	100	100.0	0.0	8-18
3/4"	100	100	1.7	100	100	100.0	0.0	8-18
1/2"	100	100	1.2	57.6	100	100.0	0.0	8-18
3/8"	100	100	1.2	30.4	90	94.9	5.1	8-18
4	98.9	0	1.2	2.2	12	54.3	40.6	8-18/ 4-8*
8	90	0	1.1	0.5	2	44.8	9.5	8-18/ 4-8*
16	78.4	0	1.1	0.5		38.1	6.7	8-18/ 4-8*
30	62.1	0	1.1	0.5		30.2	7.9	8-18
50	25.9	0	1	0.5		12.6	17.6	8-18
100	4.3	6.1	0	0.4		2.1	10.5	1.5-5.0
FM Value	2.40	5.94	7.92	6.65	5.96	4.23		

*Lower range acceptable (for any single sieve) if aggregates are slivered, sharp or elongated.

Fig. 6.7 SCC mixture proportions design analysis for St. Lucie velocity cap

Table 6.3 SCC fresh and hardened concrete properties

Initial tests (at plant)	Mix A-5
Concrete temperature, F	67
Slump flow, inches	21
Air content, %	1.3
Unit weight, lb./cf.	149.3
Average compressive strength, psi	
24 hours	
3 days	3350
7 days	4850
28 days	7235

Case Study: Mock-Up Tests for SCC Underwater Concrete Plug – NYCDEP Emergency Access Drift Plug [2]

The Delaware Aqueduct which supplies water to New York City was constructed over 80 years ago. DEL-185 Shaft 6 rehabilitation at the Roundout-West Branch (RWB) tunnel were performed in 2009. The concrete mixture design and field mock-up tests for the NYCDEP Emergency Access Drift Plug (EADP) were performed during August–December 2009 period. Mix proportions for the self-flowing underwater EADP concrete were selected from trial mixes and tested for flowability, segregation, and strength and for compliance with the project requirements. The concrete trial mixes were performed at the Dutchess Quarry Plant, while the mock-up tests were performed at the Shaft No. 6 site.

Concrete Design Mix Requirements The concrete materials and design mix for the EADP concrete, dispensing sequence for the concrete admixtures including anti-washout admixtures were selected for optimizing concrete flowability. Placement procedures for concrete including bladder valve in slick line to minimize segregation during pumping, temperature rise in concrete after placement were established for successful production and pumping of concrete for the under-water applications.

The major requirements for the SCC were:

Strength f'c: 4000 psi (trial batch strength = 5200 psi)
Flowability: 22–30 inches per ASTM C1611 to meet self-leveling without consolidation
Air Content: As required for workability and pumping of fresh concrete (no freeze-thaw durability requirement)
Maximum Water/Cementitious Material Ratio: 0.40
Minimum Cementitious Material: 570 lbs./cu yd.

To reduce the heat of hydration, a Type II cement (ASTM C150) with maximum tricalcium silicates and tricalcium aluminates not exceeding 58% and mineral admixtures, including ground-granulated blast-furnace (GBBF) slag (ASTM C989, Grade 100 or 120), or fly ash (ASTM C618, Class F) were specified.

Cement: The cement was furnished from the Lafarge Plant located in Ravena, New York (ASTM C150 Type I-II, with low alkalis and C_3A content).
Fly ash: Class F fly ash (ASTM C618 Class F) was furnished from Brayton Point source (Georgia). The fly ash had low SO_3 and loss on ignition.
GGBF: Slag cement, conforming to ASTM C989, Grade 100 or 120, was furnished from Lafarge NewCem Plant.
Coarse aggregate: ASTM C33, Size #7 (1/2″ to No. 4) or Size #8 (3/8″ to No. 4).
Fine aggregate: ASTM C33 sand.

Admixtures: All admixtures were provided by BASF. Catalog cuts for BASF admixtures, including MBVR, Delvo, HRWRA (Glenium 7500), and AWA (UW-450). To provide viscosity-modifying concrete and to minimize washout during exposure to water, an anti-washout admixture (AWA) was also specified. The AWA was added with a high-range water-reducing admixture (HRWRA) to maintain flowability.

6.3 Concrete Trial Mixtures

Preliminary trial mixtures for the SCC are shown in Table 6.4.

Concrete mixes incorporating fly ash was performed with 25% and 35% of cement replacement by weight, to verify concrete flowability, segregation, and strength characteristics. Concrete mixes incorporating GGBF slag was performed with 70% cement replacement.

Table 6.4 Preliminary concrete trial mixes (per cubic yard)

Designation	Mix 1 with 70% slag	Mix 2 with 35% fly ash	Selected mix with 25% fly ash
Water/cementitious material ratio	0.44	0.44	0.44
Cement Type I-II, lb.	197	428	544
Slag or fly ash, lb.	Slag 461	Fly ash 230	Fly ash 181
Water, lb.	289	283	319
Sand, SSD, lb.	1534	1551	1415
#7 stone, SSD, lb. (DQ)	1383	1352	1350
MBVR, oz./100 cwt	2.5	2.00	3.00
Delvo, oz./100 cwt	2.0	3.0	3.0
HRWRA, oz./100 cwt (see note)	13.0	13.0	20.0
AWA, oz./100 cwt	12.00	10.00	5.00
Concrete temperature, F	80	70	73
Initial slump (S)/flow (F), inches	28.75 (F)	27.25 (F)	9.50(S)
Air content, %	2.0	4.8	8.1
Unit weight, pcf	148.2	143.6	137.0
After 45 minutes (HRWRA+ AWA)			
Flow, inches	No reading	16.25	27.00
Air content, %		2.2	1.6
Unit weight, lb./cu		151.3	148.0
Cylinder compressive strength, psi			
1 day	539	1684	1439
3 days	2530	3066	2928
7 days	4315	4225	4107

Note: The test data indicated that for improving flowability of the AWA trial mixes, the HRWRA (75% of total dosage) should be added, along with the AWA, 45 minutes after the introduction of water to the batch

The anti-washout admixture (AWA), due to its thixotropic characteristics, prevents the dilution and segregation of concrete and impacts the flowability and self-leveling of concrete required for underwater placements. Therefore, in order to improve the flowability, a HRWRA is required. The dosage rates and the dispensing sequence, of the AWA and HRWRA, were varied in the trial mixes. The dosage rate of AWA is established from the washout test in accordance with CRD C61. An acceptable value of the loss of paste in a washout test is 5% (Figs. 6.8, 6.9, 6.10 and 6.11).

Fig. 6.8 CRDC washout test bucket

Fig. 6.9 Washout test (CRDC-61) for SCC

Fig. 6.10 Typical slump flow test: 27.25 inches spread

Fig. 6.11 Making 4 × 8 inches strength cylinders

Based on the estimated concrete delivery time from the plant to the site, HRWA and AWA were added to all the concrete trial batches 45 minutes after the initial introduction of water.

Three trial mixes were made in a 2.7 cubic feet batch using a portable mixer, in accordance with ACI 211.1. Plastic concrete tests were performed on each trial mix and included slump (ASTM C143) or flowability (ASTM C1611), air content (ASTM C231), ambient air and concrete temperatures (ASTM C1064), and unit weight (ASTM C138) measurements. 4 inches by 8 inches cylindrical specimens were cast, in accordance with ASTM C192, and subsequently tested for compressive strength. The specimens were water cured until tested at 1, 3, 7, and 28 days after casting.

Mix 1: Initial trial batches with GGBF slag cement exhibited excellent initial flowability (28.75 inches). The flowability was significantly reduced (16.75 inches) after 45 minutes. After the addition of AWA, further stiffening was observed, and no flowability reading could be taken. As a result, further testing of GGBF slag mixes was discontinued.

Mix 2 and 3: Initial trial batches with fly ash exhibited excellent initial flowability (27.25–28.75 inches), which was maintained for 45 minutes. The flowability after the addition of the AWA was 16.25 inches, less than the minimum required flowability (22 inches): however, the flowability results were better than the slag mixes. In addition, the 1-day compressive strength for the fly ash mixes was twice that of the slag mixes. Based on additional testing, the selected concrete mix proportions, containing 25% fly ash, which exhibited excellent flowability (27.25 inches) and 1-day compressive strength, were used for the EADP concrete mock-up tests.

Table 6.4 and Figs. 6.12 and 6.13 present test results summary for the preliminary trial mixes, consisting of varying percentages of GBBS slag and fly ash, including the flowability and strength data.

6.4 Preliminary Mock-Up Tests

The results of mock-up tests are presented in Table 6.5.

The layout for the mock-up tests included a 500-ft long, 5 inches diameter, slickline, which was set up on a hill slope, with approximately 150-ft height drop from

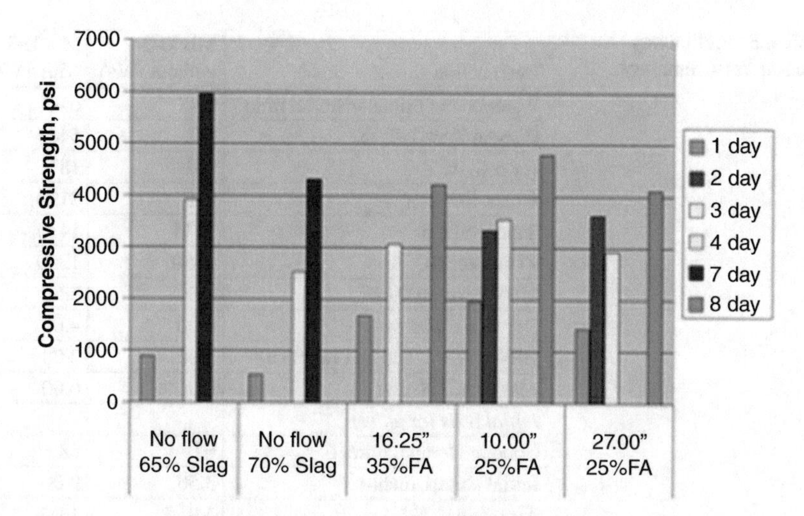

Fig. 6.12 Initial trial mixtures with GGBF slag and fly ash flowability and compressive strength

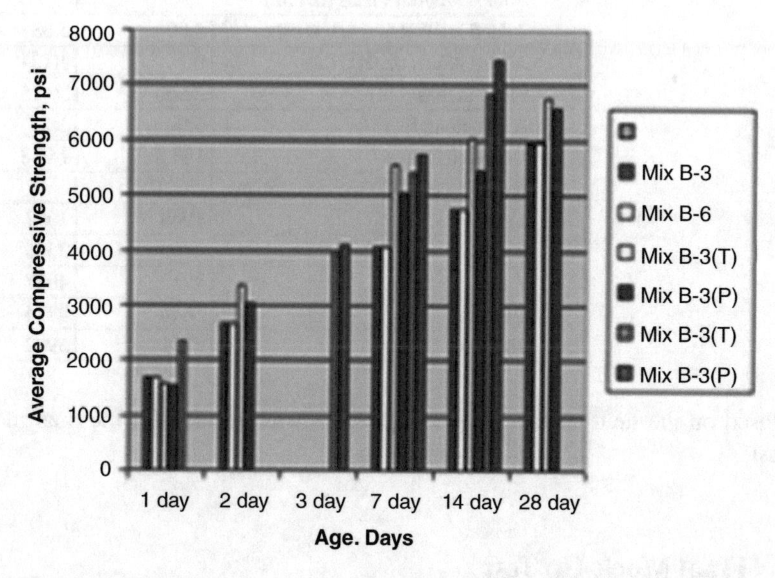

Fig. 6.13 Average compressive strength for mixes B-3 and B-6

the truck discharge elevations. At the bottom of the hill slope, the slickline was connected to 4 × 4 × 8 ft. high forms, that were placed in a sealed water-filled tank (con-ex), to receive concrete at the bottom. The tank and the forms were filled with water. The concrete was discharged into the bottom of the form.

Prior to pumping, the slickline was primed with approximately 3 cy grout; the grout was emptied into a lined waste basin. Once concrete appeared at the end of the slickline, the y-valve diverted the flow into the forms. Upon filling the form, the y-valve diverted the remaining concrete into the waste basin. Concrete samples were also taken from this location for testing, including casting strength cylinders.

Table 6.5 Preliminary mock-up tests summary

Designation	Mix B-3 without AWA	Mix B-3 with AWA
Water/cementitious material ratio	0.43	0.43
Cement Type I-II, lb.	544	544
Fly ash, lb.	181	181
Water, lb.	319	319
Sand, SSD, lb.	1415	1415
#7 stone, SSD, lb.	1350	1350
MBVR, oz./100 cwt	1.00	3.50
Delvo, oz./100 cwt	4.00	4.00
HRWRA, oz./100 cwt (at plant)	3.00	3.00
AWA, oz./100 cwt	0.00	0.00
Initial tests (at plant)		
Concrete temperature, F	67	68
Initial slump, inches	3.50	9.00
Air content, %	4.8	13.0
Unit weight, lb./cu	144.6	126.1
After 45 minutes tests (at site)		
Added HRWRA, oz./100 cwt	5.00	5.88
Added AWA, oz./100 cwt	0	10.24
Flow, inches	28.50	30.25
Air content, %	5.0	1.1
Unit weight, pcf	144.6	146.8
Average compressive strength, psi		
1 day	1940	1697
2 days	2790	2666
7 days	4210	4037
14 days	5046	4754
28 days	5761	5972

Based on the field tests, Mix B-3 with AWA was selected for the final mock-up test.

6.5 Final Mock-Up Test

A final mock-up test for underwater concrete placement with a 4-inch slickline setup was similar to the preliminary mock-up test, except a 72-inch diameter × 21-feet-long precast concrete pipe was used to represent the actual length that was used as a form. The pipe form was provided with vent pipes, Fuko hoses for secondary grouting, and was placed in a 40-feet-long water tank. Stainless steel cooling pipes for temperature control and temperature sensors were installed in the pipe, for controlling and monitoring temperatures in concrete during curing, which was also filled from the bottom, requiring 16 cy concrete.

Mix B-3 was batched at the central plant. The slickline, which was primed with 1:1 cement water slurry, instead of sand-cement grout, and the use of a 10-inch long rubber pig was inserted to separate the grout from the concrete. Tests for slump, air, and unit weight were taken for each batch, and cylinders were cast for testing strength at 3-, 7-, and 28-day ages. The concrete pipe was sliced in sections after 7 days for verification of filling and voids. Figure 6.14 shows the properties of SCC before and after the addition of AWA.

6.6 Results

A total of 15 cores, taken from the concrete block, were tested for compressive strength in accordance with ASTM C42. The average compressive strength was 4883 psi at 16 days age.

Table 6.6 presents a summary of the compressive strength test results for Mix B-3.

The thermo-couple data for Mix B-3 and Mix B-6 was analyzed to determine the compressive strength of each mock-up test block using the maturity method, in accordance with ASTM. Figure 6.15 presents the compressive strength for Mix B-3 and Mix B-6 by the maturity method.

Case Study: Chickamauga Lock Replacement, Chattanooga, TN (2019)

Chickamauga Lock Replacement is under construction on the Tennessee River in Chattanooga, TN. For construction of the drilled shafts, a SCC mixture was successfully tested. The SCC mixture proportions are shown in Table 6.7 and Fig. 6.16. The SCC mixture was tested for washout test in accordance with CRD C-61.

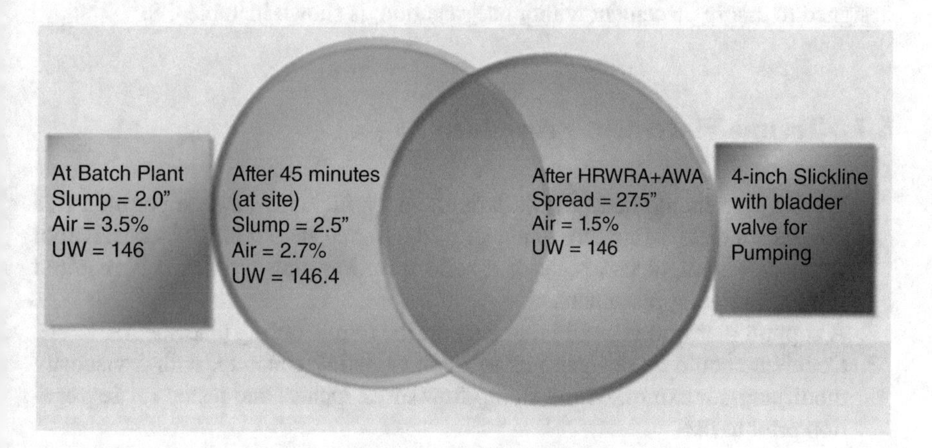

Fig. 6.14 SCC fresh concrete test results before and after the addition of AWA

Table 6.6 Final mock-up test summary

	Temp	Slump/ spread	Air content	Unit weight	Compressive strength, psi			
	Deg F	Inches	%	Lb./cf	1 day	3 days	7 days	28 days
At truck discharge (1)	81	14.75	1.5	146.4	2373	4397	5980	7090
At truck discharge (2)	76	27.5	1.5	146		3563	4817	6560
At pump discharge (3)	72	28.5	1.2	147.8		4100	5730	7453
Average								

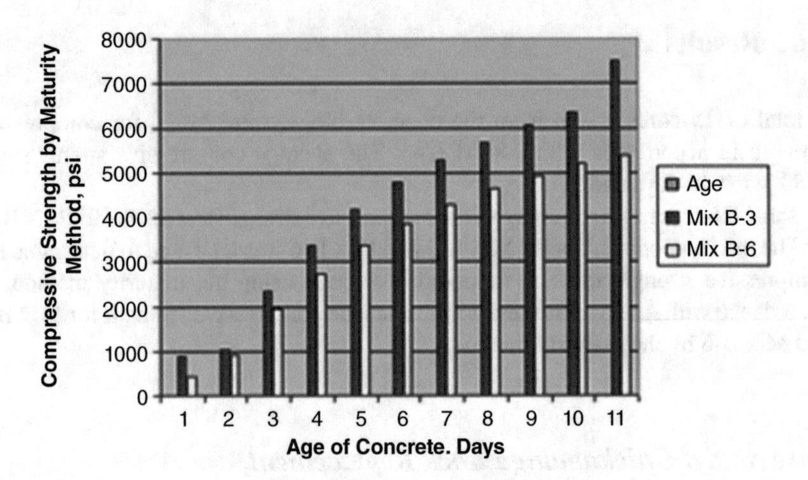

Fig. 6.15 Compressive strength test results (maturity method)

Case Study: Lower Granite Dam SCC Mixture for Intake, Idaho (2010)

Underwater concrete placement by tremie method for permanent closure of fish screen slot at Lower Granite Dam was made in 2015. The tremie concrete mixture, designed to assure placement without segregation, is shown in Table 6.8.

6.7 Tremie Placement Procedure

1. Tremie pipe should be rigid, watertight, and sufficiently large to permit a free flow of concrete. Minimum diameter of the tremie should be 10 inches. A foot valve or watertight end plate or a solid traveling plug (go-devil) is required prior to initiating placement.
2. A support platform is required to support the tremie during placement.
3. Concrete should be designed as self-consolidating concrete, with a viscosity modifying admixture, with a slump flow of 22 inches, and tested for segregation prior to use.

Table 6.7 SCC mixture proportions with #67 stone for drilled shaft (tremie concrete)

Designation	Quantities
Water/cementitious material ratio	0.42
Cement Type I-II, lb./cy	487
Slag 1 grade 100, lb./cy	0
Fly ash Class F, lb./cy	263
Water, lb./cy	315
Sand, SSD, lb./cy	1412
#67 stone, SSD, lb./cy (kg/m^3)	1499
AEA, oz./ yd	0
Plastol 6420, oz./ yd	120
AWA, oz./yd.	40
Tests after batching (at plant)	
Description	Average (15 tests)
Concrete temperature, F (C)	67
Initial slump, inches (mm)	10
Air content, %	4.8
Unit weight, lb./cf	147.8
Initial set, hours	6.5
Final set, hours	8.75
Bleed, %	0
Wash-out, %	3.4
Compressive strength, psi	
3-day	2810
7-day	4158
28-day	6056

4. Concrete may be delivered to the tremie using a concrete pump; however, the pump line should discharge into a hopper, attached to the tremie pipe.
5. The tremie bottom should be placed minimum 6 inches from the bottom to allow water and concrete to escape. Tremie should have markings at 1-ft interval to verify volume of placement and raising points
6. Place a hopper above the tremie pipe. The hopper should have adequate capacity (0.3–0.5 cy) for uninterrupted flow without segregation of concrete.
7. If a tremie pipe does not have an end plate, insert a go-devil in the pipe to keep water from penetrating the first concrete placement in the pipe.
8. Add concrete slowly and continuously to assure pipe is kept full of concrete.
9. Lift tremie after concrete has been brought to the required minimum height – depth of 2–3 feet.
10. Calculate the concrete placement per tremie location. The tremie pipe placement should limit the horizontal flow of concrete to 15 feet maximum. For example, if a placement depth is 8 ft., assuming a concrete spread of 12 feet, it will require 4 cy/ft. depth or 32 cy for the full height.
11. Monitor placement rate with drop lines to verify actual depth. If there is a delay in concrete placement greater than 30 minutes, tremie pipe should be pulled and resealed.
12. Do not move the tremie horizontally during a placing operation.

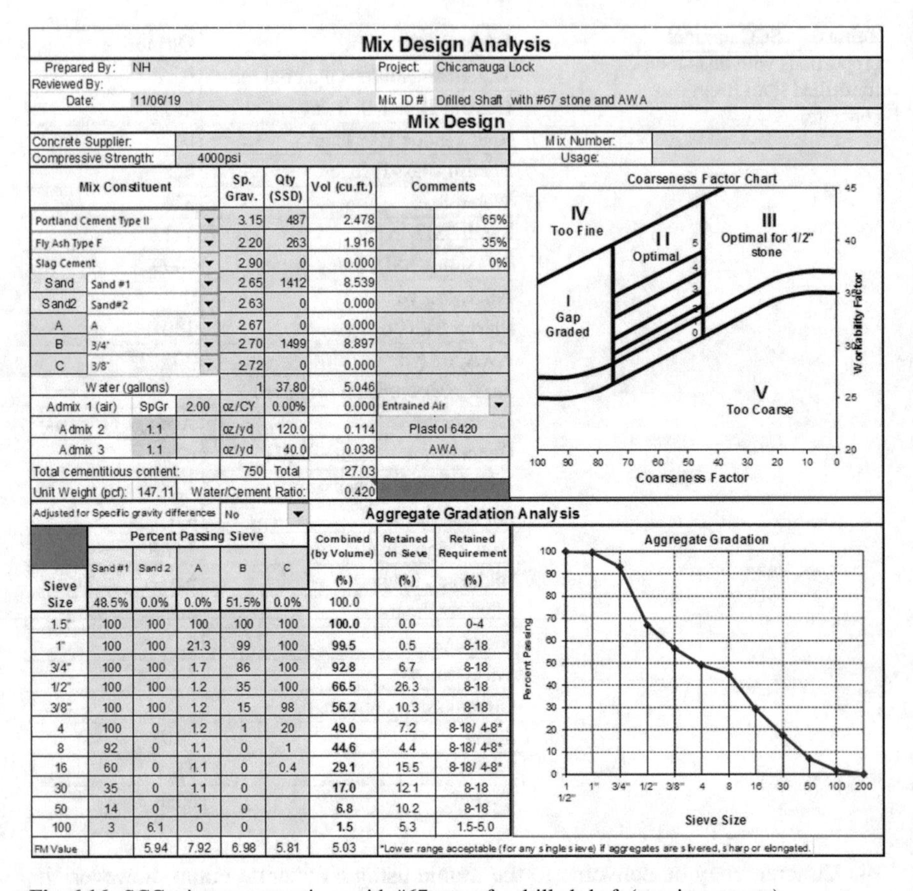

Mix Design Analysis

Prepared By:	NH	Project:	Chicamauga Lock
Reviewed By:			
Date:	11/06/19	Mix ID #	Drilled Shaft with #67 stone and AWA

Mix Design

Concrete Supplier:		Mix Number:	
Compressive Strength:	4000psi	Usage:	

Mix Constituent				Sp. Grav.	Qty (SSD)	Vol (cu.ft.)	Comments
Portland Cement Type II				3.15	487	2.478	65%
Fly Ash Type F				2.20	263	1.916	35%
Slag Cement				2.90	0	0.000	0%
Sand	Sand #1			2.65	1412	8.539	
Sand2	Sand#2			2.63	0	0.000	
A	A			2.67	0	0.000	
B	3/4"			2.70	1499	8.897	
C	3/8"			2.72	0	0.000	
Water (gallons)				1	37.80	5.046	
Admix 1 (air)	SpGr	2.00	oz/CY	0.00%		0.000	Entrained Air
Admix 2	1.1		oz/yd	120.0		0.114	Plastol 6420
Admix 3	1.1		oz/yd	40.0		0.038	AWA
Total cementitious content:			750	Total		27.03	
Unit Weight (pcf):	147.11	Water/Cement Ratio:				0.420	

Adjusted for Specific gravity differences: No

Aggregate Gradation Analysis

Sieve Size	Sand #1	Sand 2	A	B	C	Combined (by Volume) (%)	Retained on Sieve (%)	Retained Requirement (%)
	48.5%	0.0%	0.0%	51.5%	0.0%	100.0		
1.5"	100	100	100	100	100	100.0	0.0	0-4
1"	100	100	21.3	99	100	99.5	0.5	8-18
3/4"	100	100	1.7	86	100	92.8	6.7	8-18
1/2"	100	100	1.2	35	100	66.5	26.3	8-18
3/8"	100	100	1.2	15	98	56.2	10.3	8-18
4	100	0	1.2	1	20	49.0	7.2	8-18/ 4-8*
8	92	0	1.1	0	1	44.6	4.4	8-18/ 4-8*
16	60	0	1.1	0	0.4	29.1	15.5	8-18/ 4-8*
30	35	0	1.1	0		17.0	12.1	8-18
50	14	0	1	0		6.8	10.2	8-18
100	3	6.1	0	0		1.5	5.3	1.5-5.0
FM Value		5.94	7.92	6.98	5.81	5.03		

*Lower range acceptable (for any single sieve) if aggregates are slivered, sharp or elongated.

Fig. 6.16 SCC mixture proportions with #67 stone for drilled shaft (tremie concrete)

13. After concrete has been placed to the required depth, say 4 ft (16 cy), the lift tremie pipe may be raised by 18–24 inches, such that the tremie maintains a 2.0 ft minimum embedment.
14. If a loss of seal occurs, evident by increased flow rate in the tremie, follow the initial reseal procedure prior to continuing.

6.8 Conclusions and Recommendations

Based on the above case studies for SCC mixtures, including underwater placements, the following findings and recommendations are made:

1. SCC mixtures offer satisfactory workability, strength, and durability and can be prepared with locally available materials. The use of SCC is increasingly being considered for cast-in-place concrete applications for infrastructure projects, requiring high-performance concrete with improved mechanical properties.

Table 6.8 Mixture proportions for SCC placement using tremie

Designation	Granite Dam (2010)
Water/cementitious material ratio	
Cement Type I, lb./cy	468
Silica fume, lb./cy	25
Fly ash Class F, lb./cy	212
Water, lb./cy	265
Sand, SSD, lb./cy	1480
#8 stone, SSD, lb./cy (kg/m³)	1510
AEA, oz./ yd	
HRWRA, gal./ yd	72 oz./cwt
AWA, oz./yd.	26 oz./cwt
Tests after batching	
Description	
Concrete temperature, F (C)	70
Initial slump, inches (mm)	21
Air content, %	1,3
Unit weight, lb./cf	149.3
Initial set, hours	6.5
Final set, hours	8.75
Bleed, %	0
Washout, %	3.4
Compressive strength, psi	
3 days	3350
7 days	4850
28 days	7237

2. For successful SCC applications, the concrete mix must be carefully selected through laboratory and field tests to assure compliance with the anticipated the workability, slump retention, flowability, and strength requirements. Local availability of the concrete components, including fly ash and admixtures for use in SCC mixtures, should also be considered in selecting the appropriate concrete mix design.
3. The SCC mixture, incorporating Type I/II cement, fly ash, HRWRA, and VMA (AWA), requires a slump flow exceeding 25 inches (650 mm) for completely filling underwater forms with minimal segregation and voids, requiring little subsequent grouting. The compressive strength for the SCC mixtures, including 25% fly ash as partial replacement of cement and AWA should exceed 5000 psi (34.5 MPa) at 28 days.
4. The incorporation of the HRWRA and AWA in concrete provides the necessary flowability and mobility to move underwater with little washout and also enhances the in situ compressive strength. For SCC concrete, the flowability is enhanced when the VMA is dispensed to concrete batch at the site, preceded by HRWRA. The actual dosages of the HRWRA and VMA, including timing and dispensing sequence of the AWA, should be established by field testing based on the specific site conditions.

5. The compatibility between the HRWRA and AWA, including dosage requirements, and mixing time should also be established by site-specific testing.
6. Placing of SCC requires a watertight formwork and adequate formwork design to sustain the lateral hydrostatic pressures during concrete placements.
7. Tremie placements for SCC mixtures require planning and selection of appropriate equipment for successful implementation. SCC can be placed underwater concrete using a slickline with minimal bends and use of appropriate bladder and diverter valves and procedures that assure continuous pumping, continuous agitation, and control on placement temperatures that would result in sound concrete without segregation or voids.
8. Mock-up tests provide valuable information in finalizing the concrete mix design and procedures for the successful implementation of underwater concrete placement.

References

1. Hasan N, Faerman E, Berner D (1993) Advances in underwater concreting: St. Lucie plant intake velocity cap rehabilitation. In: High performance concrete in severe environment. ACI Special Publication SP-140
2. Hasan N (2012) Mock-up tests for a self-flowing underwater concrete plug. In: Tenth international conference on superplasticizers and other chemical admixtures in concrete. Czech Republic, Prague

Chapter 7
Sulfate Attack Mitigation

7.1 General

Sulfate attack can be from two sources. Internal sources of sulfate come from the materials incorporating concrete, such as mixing water, that can attack the cement paste or physical attack caused by external sulfates traveling through the concrete. The external sources of attack come from the physical environment where multiple deterioration mechanism may be involved.

ACI 201.2R-16 provides a historical perspective on sulfate attack on concrete:

> Traditionally sulfate attack has been thought to occur as a consequence of a sulfate-containing solution entering the pore structure of concrete and reacting with hydrating cement compounds such as tricalcium aluminate(C_3A) to form various sulfate-containing phases that adversely affect concrete durability

Bates [1] states "It is almost universally believed that it is the reaction of sulfate of magnesia of the sea water, with the lime of the cement, and the alumina of the aluminates in cement, , resulting in the formation of hydrated magnesia and calcium sulfo-aluminate, which crystallizes with a large number of molecules of water." This understanding has been the basis for the development of sulfate-resistant cements and mixture proportions for concretes placed in sulfate environments.

Lerch [2] established the composition of ettringite, known as a Candolt's salt 3CaO.

$Al_2O_3.3CaSO4.31H_2O$ and that of monosulfate as $3CaO.Al_2O_3.CaSO4.12H_2O$. Lerch also showed that monosulfate converts to ettringite when an external source of sulfate is provided.

Resistance to sulfate attack is increased by controlling both cement composition and concrete permeability. Verbeck [3] and Stark [4] showed that the reduction of permeability was of greater importance in limiting sulfate attack than using sulfate-resistant cement. USBR's 40-year data showed that a w/cm of 0.45 or lower helps in avoiding damage from sulfate attack on Portland cement with C_3A of less than 8 percent [5].

N. Hasan, *Durability and Sustainability of Concrete*,
https://doi.org/10.1007/978-3-030-51573-7_7

ACI 201.2R-16 also describes the mechanisms for external sulfate attack including ettringite and gypsum formation. External sulfate attack produces two damage mechanisms in concrete. Cracking due to expansion is probably the most widely reported form of damage. Expansion occurs because the volumes of ettringite and gypsum are greater than those of the reactants from which they form. An increase in volume in the solid phase in the hardened cement matrix results in tensile stresses due to crystallization pressures, and cracks develop once the tensile strength of the paste is locally exceeded. A second damage mechanism associated with external attack involves softening and loss of cohesion. The damage mechanism involves chemical alterations that destabilize the C-S-H and calcium hydroxide and can result in the formation of microcracks without significant expansion.

Refer to ACI 201.2-16 R for more information on sulfate attack. Detwiler and Powers [6] reported results of concrete with cement C_3A of 5 to 12%, and air air content of 2 to 6%, exposed to durability test by ASTM C666. Concrete performance was dominated by air content. Specimens with low air entrainment were damaged. Deposition of ettringite in concracks of marginally air-entrained concrete was found.

ACI 350R "Code Requirements for Environmental Engineering Concrete Structures" provides guidelines for concrete exposed to sulfate exposure. These are similar to ACI 318 requirements for sulfate exposure, except the compressive strength and w/cm ratios are slightly more conservative for severe and very severe sulfate exposure, as shown in Table 7.1.

For moderate exposure, Type II cement with C_3A of 8% maximum is recommended.

For concrete placed in sea, the moderate exposure requirements may be used. Even though it generally contains more than 1500 ppm SO_4, Type II cement with C_3A of 10% is allowed if the w/c ratio is reduced to 0.40.

ACI 318 has four exposure classes (S0, S1, S2, and S3) for sulfate, as shown in Tables 7.2 and 7.3.

- No requirement

Table 7.1 Requirements for concrete exposed to sulfate-containing solutions[1]

Sulfate exposure	Water-soluble sulfates (SO_4) in soil, percent by weight	Sulfate (SO_4) in water, ppm	Cementitious material[2]	Minimum strength f'c (psi)	Maximum water/cementitious material ratio
Negligible	00–0.10	0–150	No restriction		0.45
Moderate	0.10–0.20	150–1500	Type II, IP (MS), IS (PM)(MS), I(SM)(MS)	4000	0.42
Severe	0.20–0.2.00	1500–10,000	Type V	4500	0.40
Very severe	Over 2.00	Over 10,000	Type V cement plus SCMs	5000	0.40

[1]ACI 350–06, Table 4.3.1

Table 7.2 Concrete classes exposed to sulfates[1]

Class	Sulfate exposure	Cementitious material[2]	Minimum strength f'c (psi)	Maximum water/ cementitious material ratio
S0	$SO_4 < 150$ ppm	No restriction	2500	N/A
S1	Moderate ($150 < SO_4 < 1500$ ppm)	Use Type II cement or blended cement	4000	0.45
S2	Severe ($150 < SO_4 < 10{,}000$ ppm)	Use Type V cement or blended cement	4500	0.45
S3	Very severe ($SO_4 > 10{,}000$ ppm)	Use Type V cement and SCMs	4500	0.45

[1]Refer to ACI 318 Building Code
[2]Alternative combinations of cementitious materials, including natural pozzolans, silica fume, and slag, may be used to improve sulfate resistance of concrete, based on testing

Results The study concluded that concrete mixtures with 28-day design compressive strength of 4500 psi and a w/cm of 0.45 or less are required to provide high resistance to PSA. Supplementary cementitious materials (SCMs) do not improve the resistance of concrete to PSA. They did not lead to poor performance either.

Case Study: Tertiary Treatment Facility (TTF), Sacramento

The Sacramento Regional County Sanitation District (Regional San) operates the Sacramento Regional Wastewater Treatment Plant (SRWTP) that provides wastewater treatment to the Sacramento area and surrounding cities, serving approximately1.4 million customers. SRWTP currently uses a secondary treatment process with a capacity of 181 million gallons per day (mgd) average dry weather flow (ADWF). The treated effluent is discharged to the Sacramento River.

Regional San has embarked on a comprehensive upgrade, called the EchoWater Project (Program) that includes modifications to the existing secondary process and the addition of a tertiary treatment processes. The tertiary improvements include filtration and enhanced disinfection, consistent with recycled water requirements under Title 22, Division 4, Chap. 3 of the California Code of Regulations or equivalent. The tertiary treatment process will provide up to 217 mgd of effluent.

Begun in 2011, the Program has a 12-year schedule and an estimated construction cost of $1.735 billion. The Program is focused on meeting effluent discharge requirements well into the future.

The two areas of focus for this paper are the structures for the Nitrifying Sidestream Treatment (NST) and the Tertiary Treatment Facilities (TTF); see Fig. 7.1. The NST Project concrete is discussed in Chap. 8.

TTF Concrete Mixtures for Durability The concrete structures are designed in accordance with ACI 350 Code applicable to environmentally exposed structures, with a maximum w/cm ratio of 0.40 and design compressive strength of 4500 psi at

Table 7.3 ACI 201.2 requirements to protect concrete exposed to sulfate attack[1]

		Cementitious material requirements			Performance requirements maximum expansion		
Class	Maximum water/cementitious material ratio	ASTM C150/ C150M	ASTM C595/ C595M	ASTM C1157/ C1157M	ASTM C1012/C1012M		
					6 months	12 months	18 months
S0	No restriction	No restriction	None	None			
S1	0.50	Type II	IP(MS),				
			IS (<70) (MS)				
			IT(P < S < 70)MS	MS	0.1%	*	*
			IT (P > S) (MS)				
S2	0.45	Type V	IP(MS),				
			IS (<70) (MS)				*
			IT(P < S < 70) MS	HS	0.05%	0.1%	
			IT (P > S) (MS)				
S3	0.45	Type V plus SCMs	IP(MS),				
			IS (<70) (MS)			*	0.1%
			IT(P < S < 70)MS	HS	*		
			IT (P > S) (MS)				

[1]Refer to ACI 201.2R-16, Table 6.1.4.1.b

[2]Alternative combinations of cementitious materials, including natural pozzolans, silica fume, and slag, may be used to improve sulfate resistance of concrete, based on testing

Fig. 7.1 Proposed Tertiary Treatment Facility (TTF), Sacramento. (Courtesy of TTF EchoWater Project)

Table 7.4 Specified performance requirements for Class 1 structural concrete

			With MasterLife 300D (BASF)
Permeability test	CRD-48 DIN 1048 (96 hour)	No Leakage	5.24 × 10^ – 11 m/s No penetration
Sulfuric acid resistance	7% sulfuric acid for 40 days	40% max. mass loss	30% loss in 1.0 pH solution
Sulfate resistance	ASTM C1012	< 0.1% expansion	0.085% in 1 year
Chloride diffusion	ASTM C1556	25% min reduction	>25%
Compressive strength	ASTM C39	No reduction in strength	None

28 days. Additional performance requirements for permeability, sulfate resistance, and chloride resistance for the liquid retaining TTF structures are shown in Table 7.4.

For compliance with the above performance requirements for Class 1 concrete, trial mixtures included a shrinkage-reducing admixture and a crystalline capillary waterproofing admixture (PRA) to reduce permeability and prohibit penetration of water and other liquids.

Shrinkage-reducing admixtures (MasterLife SRA 035) reduce surface tension of the pore solution by a factor of 2, resulting in decrease in capillary stress (reduction in microstrain from 200 microstrain to 100 microstrain at 400 hours after the hydration is commenced).

MasterLife 300D is a crystalline capillary waterproofing admixture, from BASF Corporation, which meets the performance requirements shown in Table 7.5.

Table 7.5 Typical data for MasterLife 300D @ 2% by mass of cement

Performance characteristics	Test method	Performance relative to control concrete mixture
Capillary absorption	ASTM C1585	43% reduction
Water penetration	Modified DIN 1048	40% reduction
Moisture vapor emission rate	ASTM F 1869	40% reduction
Electrical conductance	ASTM C1202	40% reduction
Compressive strength	ASTM C39	7% increase

Table 7.6 High-performance concrete mixture proportions

Designation	Quantities
Water/cement ratio	0.40
Cement type I–II, Lehigh, lb./yd^3, (kg/m^3)	320 (190)
Slag grade 100/120, lehigh lb./yd^3(kg/m^3)	320 (190)
Fly ash class F, lb./yd^3(kg/m^3)	0
Water, lb./cy(kg/m^3)	258 (153)
Sand, SSD, granite rock lb./yd^3(kg/m^3)	1333 (791)
#57 stone, SSD, teichert, lb./yd^3(kg/m^3)	1562 (927)
#8 stone, SSD, lb./yd^3(kg/m^3)	180 (107)
Master air AE200, oz./yd^3(ml/m^3)	15.0 (580)
MasterPolyheed 997, fl. oz./yd^3(ml/m^3)	64 (2475)
MasterLife SRA 035, fl. oz./yd^3 (ml/m^3)	128 (4950)
MasterLife 300 D, lb./yd. ^{3} (kg/m^3)	6.5 (7.7)
Initial tests after batching (at plant)	
Concrete temperature, F (C)	67 (19.4)
Initial slump, inches (mm)	4.0 (100)
Air content, %	4.0
Density, lb./cf., (kg/m^3)	147.2 (2358)

Concrete mixture proportions for Class 1 concrete (4500 psi), being used for the TTF project, are shown in Table 7.6 and Fig. 7.2.

Additional data on compressive strength is included in Chap. 9.

Case Study: Reference to SP 317-7 – Criteria for Selecting Mixtures Resistant to Salt Attack (By Karthik H Kobla and Robert C. O'Neill [7])

The above paper presents the results of a recent 2014 study on the resistance of concrete to physical salt attack (PSA) by exposing concrete prisms in sulfate solution. Several concrete mixtures were made with varying w/cm (0.40–0.60), cement

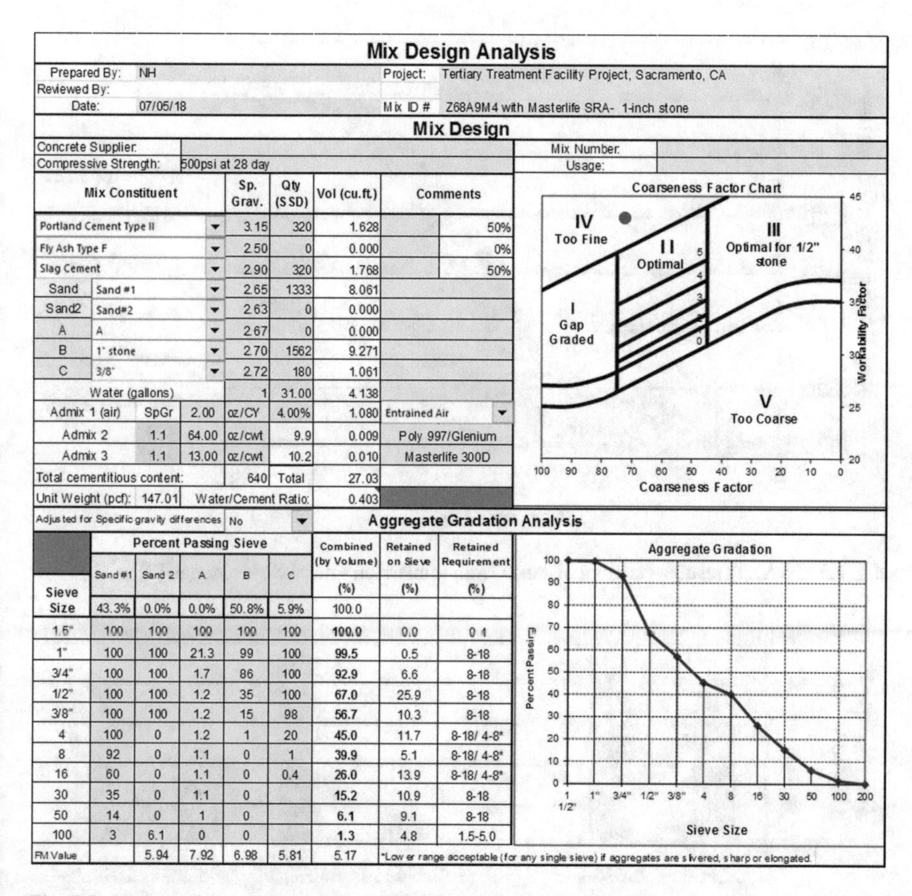

Mix Design Analysis

Prepared By:	NH	Project:	Tertiary Treatment Facility Project, Sacramento, CA
Reviewed By:			
Date:	07/05/18	Mix ID #	Z68A9M4 with Masterlife SRA- 1-inch stone

Mix Design

Concrete Supplier:	
Compressive Strength:	500psi at 28 day
Mix Number:	
Usage:	

Mix Constituent		Sp. Grav.	Qty (SSD)	Vol (cu.ft.)	Comments
Portland Cement Type II		3.15	320	1.628	50%
Fly Ash Type F		2.50	0	0.000	0%
Slag Cement		2.90	320	1.768	50%
Sand	Sand #1	2.65	1333	8.061	
Sand2	Sand#2	2.63	0	0.000	
A	A	2.67	0	0.000	
B	1" stone	2.70	1562	9.271	
C	3/8"	2.72	180	1.061	
Water (gallons)		1	31.00	4.138	

Admixture	SpGr		Unit	Qty	Vol (cu.ft.)	Comments
Admix 1 (air)	SpGr	2.00	oz/CY	4.00%	1.080	Entrained Air
Admix 2	1.1	64.00	oz/cwt	9.9	0.009	Poly 997/Glenium
Admix 3	1.1	13.00	oz/cwt	10.2	0.010	Masterlife 300D

Total cementitious content:	640	Total	27.03
Unit Weight (pcf): 147.01		Water/Cement Ratio:	0.403
Adjusted for Specific gravity differences	No		

Aggregate Gradation Analysis

Sieve Size	Percent Passing Sieve: Sand #1	Sand 2	A	B	C	Combined (by Volume) (%)	Retained on Sieve (%)	Retained Requirement (%)
	43.3%	0.0%	0.0%	50.8%	5.9%	100.0		
1.5"	100	100	100	100	100	100.0	0.0	0 4
1"	100	100	21.3	99	100	99.5	0.5	8-18
3/4"	100	100	1.7	86	100	92.9	6.6	8-18
1/2"	100	100	1.2	35	100	67.0	25.9	8-18
3/8"	100	100	1.2	15	98	56.7	10.3	8-18
4	100	0	1.2	1	20	45.0	11.7	8-18/ 4-8*
8	92	0	1.1	0	1	39.9	5.1	8-18/ 4-8*
16	60	0	1.1	0	0.4	26.0	13.9	8-18/ 4-8*
30	35	0	1.1	0		15.2	10.9	8-18
50	14	0	1	0		6.1	9.1	8-18
100	3	6.1	0	0		1.3	4.8	1.5-5.0
FM Value		5.94	7.92	6.98	5.81	5.17	*Lower range acceptable (for any single sieve) if aggregates are slivered, sharp or elongated.	

Fig. 7.2 High-performance concrete (4500 psi) mixture design analysis

types (I, II, and V), and SCM (slag cement and fly ash combination). Concrete specimens were moist cured for cured for 28 days. The cured specimens, consisting of prisms 3 × 3 × 11 1/4 (75 × 75 × 285 mm), were immersed in 10% sodium sulfate solution after 56 days from casting. All concrete specimens were partially immersed for a period of 27 months. Damage in the form of salt crystallization and scaling was observed with time. The length of scaling front was determined above the surface of the salt solution.

The concrete specimens with Type I cements and low w/cm of 0.40 with supplementary cementitious materials (SCM), including 25% slag cement (0.40SL25) or 15% fly ash (0.4FA15), exhibited superior resistance to PSA, after 27 months of exposure, as shown in Fig. 7.3 [7].

The permeability of these specimens, subjected to ASTM C1202, rapid chloride permeability test, with standard 28-day moist curing, was 704 and 913 coulombs, respectively.

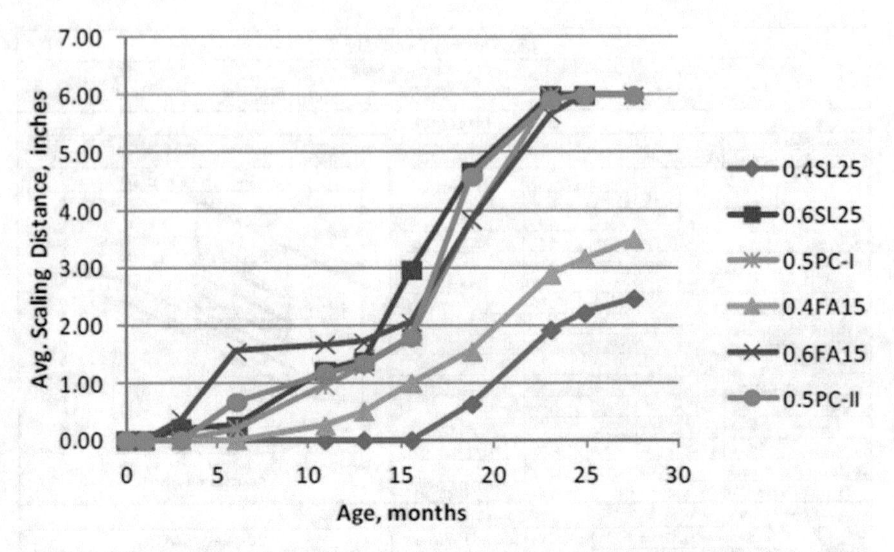

Fig. 7.3 PSA test results – scaling distance from immersion line vs exposure age [7]

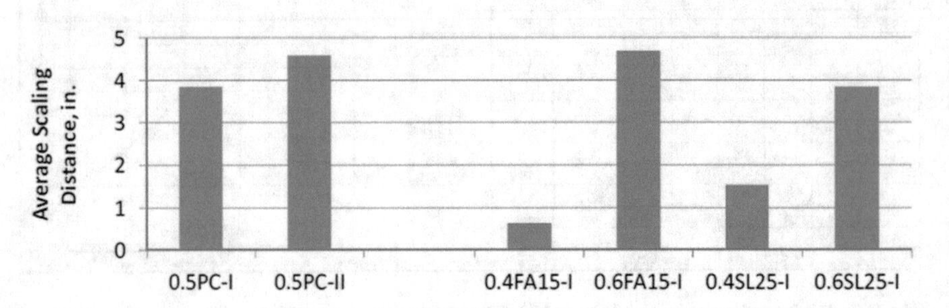

Fig. 7.4 PSA test result – 19 months of scaling distance comparison between PC and SCM mixtures

A comparison of the concrete specimens, with and without the SCM, exposed to PSA, is shown in Fig. 7.4 [7]. Specimens containing no slag cement or fly ash generally were less resistant to damage and scaling from sulfate attack after 19 months.

7.2 Recommendations

Based on the literature referenced and case studies discussed in this chapter, the following recommendations are to mitigate sulfate attack on concrete:

1. Proportion concrete mixtures to minimize the ingress and movement of water, using appropriate ingredients and admixtures and cements with low C_3A cement with lower fineness values.

2. Use a shrinkage-reducing admixture to limit cracking and a permeability-reducing admixture to reduce penetration of aggressive salts.
3. Use w/cm of 0.42 or less for concrete exposed to medium or severe sulfate exposures.
4. Perform ASTM C1012 to verify expansion characteristics.
5. To minimize the DEF reactions, limit the internal concrete temperature to 158F at any stage.
6. Avoid moisture loss during curing.

References

1. Bates PH, Phillips AJ, Wig RJ (1913) Action of Salts in Alkali Water and Sea Water on Cement. In: Technologic Papers of the Bureau of Standards, vol 12. US Department of Commerce, Washington, DC. 157 pages
2. Lerch W, Ashton FW, Bogue RH (1929) The Sulpho-aluminates of calcium, Bureau of Standards. J. Res. 2(4):715–731
3. Verbeck GJ (1958) Field and laboratory studies of the sulphates resistance of concrete. In: Thorvaldson symposium. University of Toronto Press, Toronto, ON, pp 113–124
4. Starch D (1989) Durability of concrete in sulfate-rich soils. In: Research and development bulletin RD097. Portland Cement Association, Skokie, IL. 14pp
5. Monteiro PJM, Kurtis KE (2003) Time to failure for concrete exposed to severe sulfate attack. Cem Concr Res 33(7):987–993
6. Detwiler R, Powers-Couche LJ (1999) Effect of sulfates in concrete on their resistance to freezing and thawing. In: Ettringite-the sometimes host of destruction. American Concrete Institute, SP-177, Farmington Hills, MI, pp 219–247
7. Obla KH, O'Neill RC Criteria for selecting mixtures resistant to physical salt attack. In: SP-317-7. American Concrete Institute

Chapter 8
Drying Shrinkage Mitigation

8.1 General

High-performance concrete (HPC): According to ACI, the HPC concrete is concrete meeting special combinations of performance and uniformity requirements that cannot be always achieved by routinely using conventional constituents and normal mixing, placing, and curing practices.

HPC is being increasingly used in bridge decks and pavement overlays due to its high strength, low permeability, and enhanced durability. The HPC concrete mixture has high cementitious content and a low water to cementitious ratio, of no more than 0.4, and incorporates a ternary blend using Portland cement, fly ash, and silica fume.

Shrinkage of concrete: Portland Cement Association (PCA) Publication, *Design and Control of Concrete Mixtures*, Chapter 16 [1], states:

"The shrinkage in concrete occurs in several modes: plastic, autogenous and drying shrinkage. The plastic shrinkage occurs immediately after placing of concrete due to evaporation of bleed water from the exposed surface. The autogenous shrinkage is due to volume reduction caused by the cement hydration when there is no moisture loss and occurs between initial and final set. This is an external volume reduction. Chemical shrinkage occurs due to reduction in volume in the hydrating paste. The hydration continues beyond the initial set causing the formation of voids in the microstructure. It is related to the chemistry and degree of hydration, and it is an internal reduction in volume."

Figure 8.1 shows early volume changes in concrete paste, as it hydrates at casting, at initial set, and after hardening, while Fig. 8.2 shows a relationship between autogenous shrinkage and chemical shrinkage of cement paste.

N. Hasan, *Durability and Sustainability of Concrete*,
https://doi.org/10.1007/978-3-030-51573-7_8

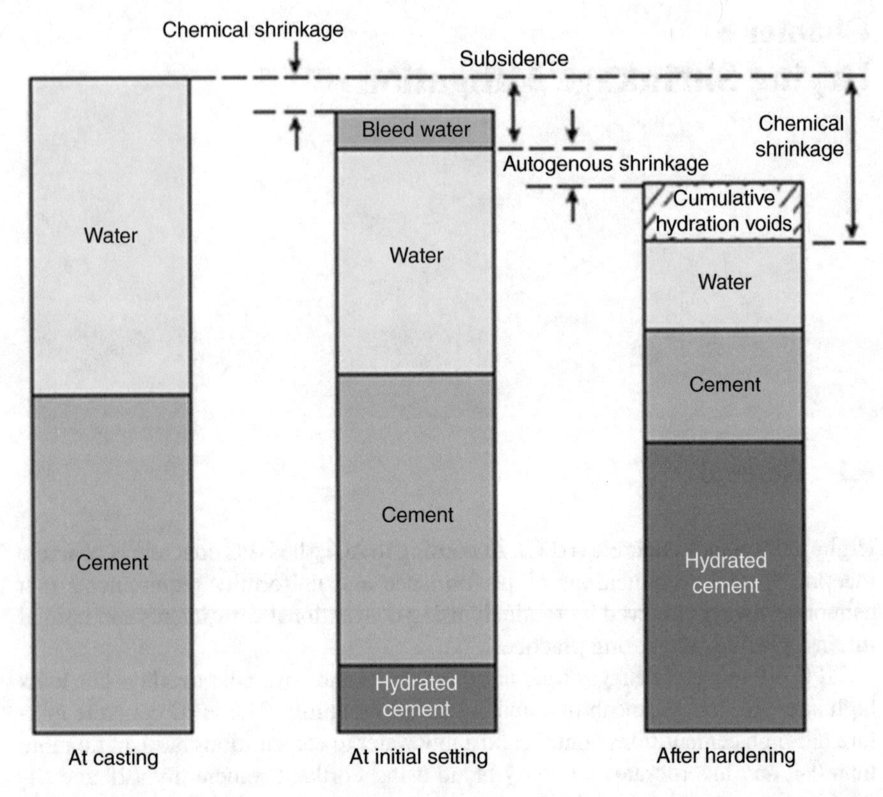

Fig. 8.1 Volumetric relationship between subsidence, bleed water, chemical shrinkage, and autogenous shrinkage. (Refer to Figure 13-5 of Ref. [1])

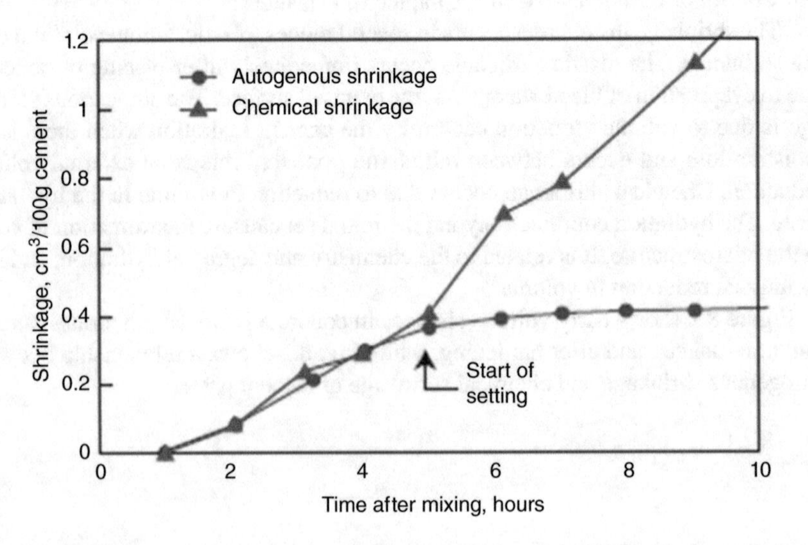

Fig. 8.2 Relation between autogenous shrinkage and chemical shrinkage of cement paste (Hammer 1999). (Refer to Figure 13-4 of Ref. [1])

8.2 Drying Shrinkage

The drying shrinkage starts after the concrete has reached final set due to reduction in volume caused from evaporation of free water in the capillary pores and continues beyond the first 7 days, due to exposure to surface temperatures and humidity. The actual mechanism of drying shrinkage is complex, but drying shrinkage is generally associated with loss of adsorbed water from the hydrated paste [2, 3].

Concrete that is completely submerged in water or in an environment with 100% humidity will not experience drying shrinkage phenomenon, as shown in Fig. 8.3. Specimen A stored in water indicated swelling, whereas Specimen B, subjected to alternate wetting and drying cycles, indicated drying shrinkage.

There are several factors that affect drying shrinkage that include concrete proportions, construction practice, and environmental conditions.

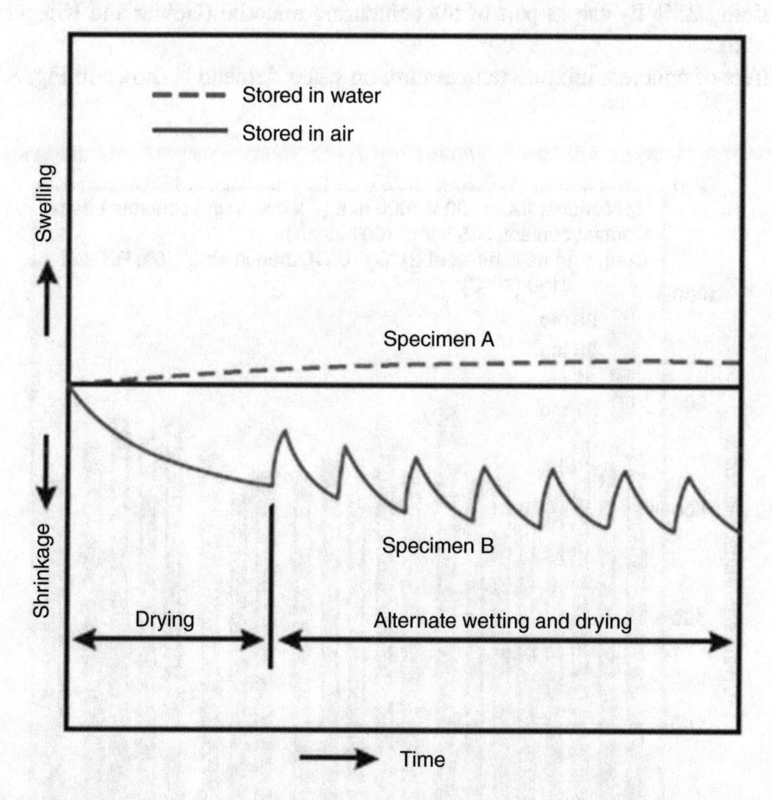

Fig. 8.3 Schematic illustration of moisture movements in concrete. If concrete is kept continuously wet, a slight expansion occurs. However, drying usually takes place, causing shrinkage. Further wetting and drying causes alternate cycles of swelling and shrinkage. (Figure 13-9 of Ref. [1])

Results of long-term drying shrinkage tests by the US Bureau of Reclamation [4, 5]: shrinkage ranged from 600 to 790 millionths after 38 months of drying are shown in Fig. 8.4. The shrinkage of concrete made with air-entraining admixture was similar to that of non-air-entrained concrete.

The most important factor affecting drying shrinkage is the amount of water in concrete. Figure 8.5 illustrates the effect of total water content on drying shrinkage of concrete, based on a large number of data [1]. The drying shrinkage of concrete can be minimized by reducing the amount of water, i.e., low w/cm and maximizing the aggregate content. The drying shrinkage is also affected by the aggregate size.

Drying shrinkage is also dependent on type of aggregate. Hard, rigid aggregates provide more restraint to drying than softer, less rigid aggregates. Refer to ACI 224R for additional information on aggregates with low drying shrinkage.

Supplementary cementitious material, such as fly ash usually has little effect on drying shrinkage, as shown in Fig. 8.7. Similar results were found with concretes containing 25% fly ash as part of the cementing material (Gebler and Klieger [6]) (Fig. 8.6).

Effect of concrete mixture temperature on water demand is shown in Fig. 8.7.

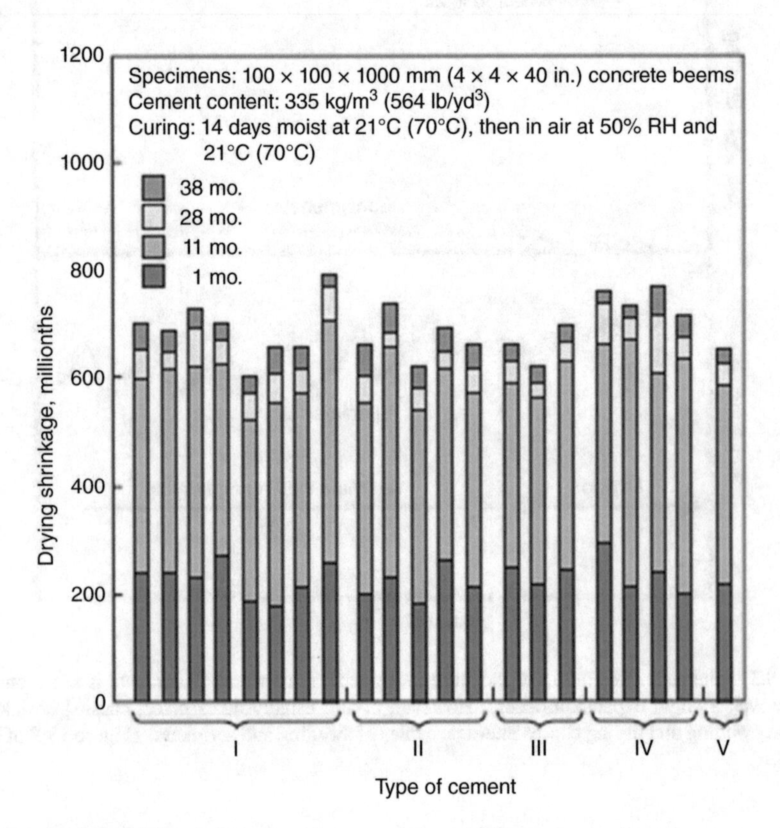

Fig. 8.4 Drying shrinkage of concrete, by type of cements [4]

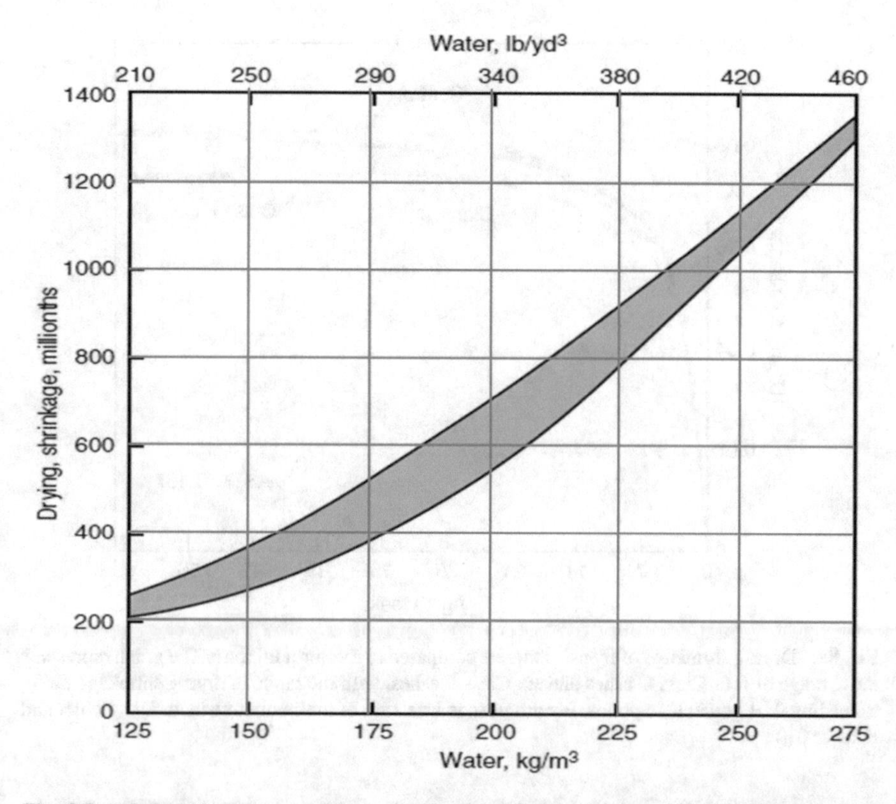

Fig. 8.5 Relationship between total water content on drying shrinkage. Shaded area represents data from a large number of mixtures of various proportions. Drying shrinkage increases with increasing water contents. (Kosmatka and Wilson [1])

Lowering the curing temperature of concrete is also effective in mitigating drying shrinkage as shown in Fig. 8.8.

One of the mechanisms of concrete cracking is volume change due to drying shrinkage. Shrinkage cracking of cast-in-place concrete structures, including walls, slabs and floors, and bridge decks, results in and corrosion of reinforcing steel, followed by spalling of concrete; it results in increased maintenance cost and reduced service life of the structure.

8.3 Free Shrinkage Test (ASTM C157)

The free drying shrinkage of concrete, determined in the laboratory by ASTM C157 test method, is expressed as a percentage or in millionths ($\times 10^6$). If the concrete is restrained from shrinkage, and the tensile stress, induced in the hydrated cement paste by the capillary forces, exceeds the local tensile strength of the concrete, cracks will occur [7].

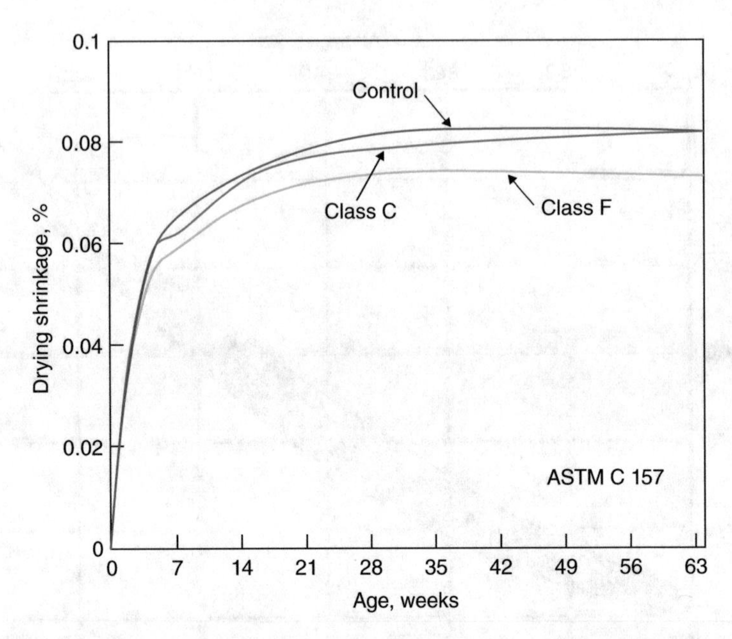

Fig. 8.6 Drying shrinkage of fly ash concrete compared to a control mixture. The graph represents the average of four Class C ashes and six Class F ashes, with the range in drying shrinkage rarely exceeding 0.01 percentage points. Fly ash dosage was 25% of the cementing material. (Gebler and Klieger [6])

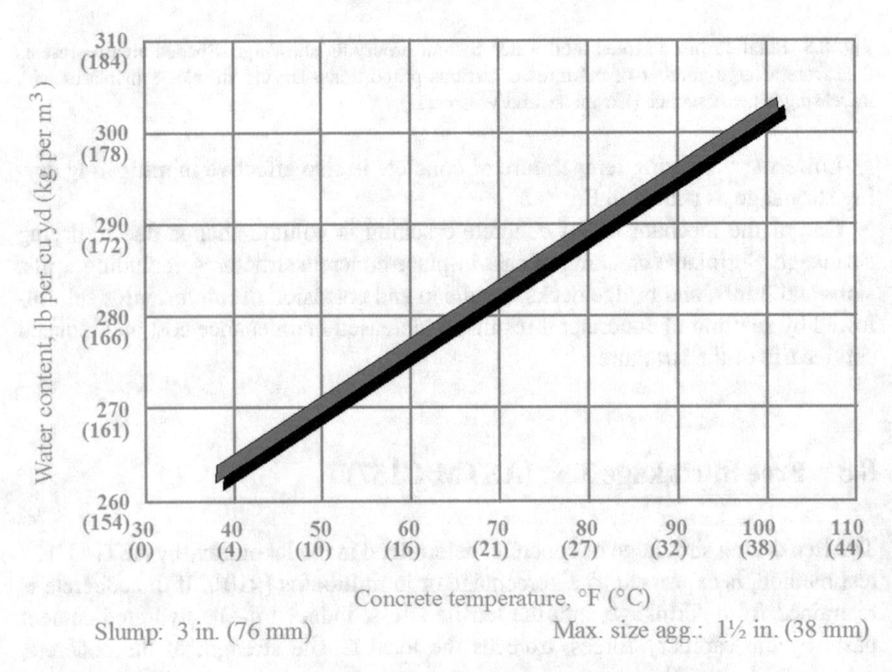

Fig. 8.7 Effect of concrete mixture temperature on water requirements [1]

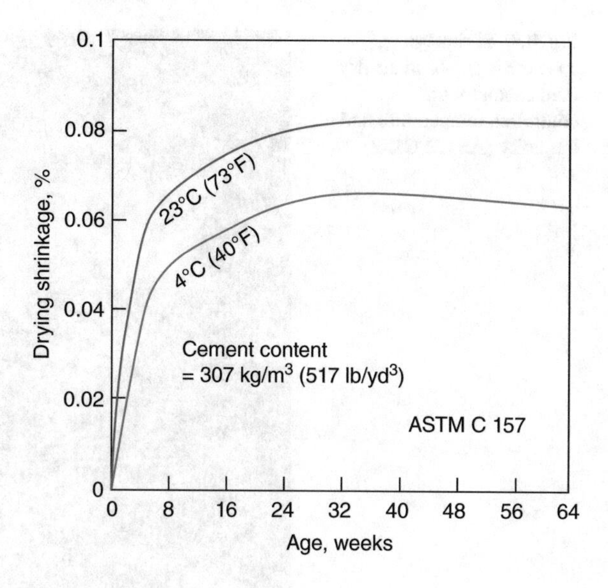

Fig. 8.8 Effect of initial curing on drying shrinkage of Portland cement concrete prisms. Concrete with an initial 7-day moist cure at 4C (40F) had less shrinkage than concrete with an initial 7-day moist cure at 23C (73F). Similar results were found with concrete containing 25% fly ash as part of the cementing material. (Gebler and Klieger [6])

Free drying shrinkage is monitored using the ASTM C157 test, which is a common method to determine length change of hardened concrete prisms 75 mm × 75 mm × 285 mm (3 × 3 × 11.25 in). The specimens were removed from the mold 24 hours after casting. Then the specimens are stored in a moist room of 23 ± 2 °C (73.5 ± 3.5 °F) and > 95% RH for the desired curing duration (i.e., 3 days, 7 days, or 14 days). Upon the end of curing duration, the specimens are moved to an environmental chamber with control drying condition of 23 ± 2 °C and 50 ± 4% RH. During drying, the length was monitored by a comparator. The mass change is also recorded.

8.4 Restrained Shrinkage Test (ASTM C1581)

ASTM C1581 test is a restrained shrinkage ring test technique to identify potential cracking of concrete.

A restrained shrinkage ring test is used as a testing technique to identify potential cracking risks of certain concrete and mortar mixtures. There are two standard testing procedures based on similar principles: ASTM C1581 and AASHTO T334 with a difference is the concrete thickness. The thickness of the concrete ring specimen for ASTM C1581 is 1.5 in, and the thickness for the AASHTO T334 ring is 3 in. ASTM test mold and test specimens are shown in Figs. 8.9 and 8.10. Based on test results, use ASTM C1581 cracking potential classification, based on stress rate, shown in Table 8.1.

Fig. 8.9 Shrinkage specimens stored in air-dry cure cabinet with controlled temperature and humidity (ASTM C157)

Fig. 8.10 ASTM test specimen mold (left) and test specimen (right)

Table 8.1 ASTM C1581 cracking potential classification (based on stress rate at time-to-cracking)

Time-to-cracking, (tcr), days	Stress rate at cracking, S, MPa/day	Potential for cracking
$0 < tcr \leq 7$	$S \geq 0.34$	High
$7 < tcr \leq 14$	$0.17 < S < 0.34$	Moderate-high
$14 < tcr \leq 28$	$0.10 < S < 0.17$	Moderate-low
$tcr > 28$	$S < 0.10$	Low

8.5 Mitigating Early-Age Cracking in HPC Bridge Decks and Pavement Overlays

The HPC concrete mixture due to its high cementitious content is prone to autogenous, plastic shrinkage and drying shrinkage, unless mitigation measures are adopted. Over the past decade, many states in the USA have reported shrinkage-related cracking in over 100,000 bridge decks, and overlays have suffered from transverse cracking, which is a pattern of drying shrinkage in concrete [8, 9].

Shrinkage cracking accelerates the deterioration of concrete and corrosion of reinforcement in concrete.

Since the volume change of concrete is a function of the cement paste, the best way to mitigate early volume change is to reduce the overall cementitious content, and by optimizing the aggregate gradation across the entire fine aggregate-coarse aggregate spectrum, such that the void space is minimized.

Studies have shown a reduction in w/cm ratio from 0.54 to 0.45; using a HRWRA will decrease the drying shrinkage 30% at 28 days.

8.6 Shrinkage-Reducing Admixtures (SRA)

Shrinkage-reducing admixtures (SRA) are typically used for mitigation of shrinkage and cracking. The SRA were first introduced in North America in 1980s. These admixtures are propylene-glycol or alkyl-ether based.

SRA will reduce the drying shrinkage and the rate of drying shrinkage of concrete. The shrinkage-reducing admixture (SRA), when added to concrete, reduces the surface tension of the water inside the capillary pores that develops during drying (Balogh 1996). As the concrete hardens, the admixture remains in the pore system and thus reduces the surface tension of the water that contributes to drying shrinkage. SRA may affect air content and compressive strength of concrete. Therefore, they are generally used in combination with water reducing admixture to reduce w/cm ratio and to improve the compressive strength and other properties of concrete.

Figure 8.11 shows a relationship of drying shrinkage of concrete with and without MasterLife SRA 20 admixture with time. The drying shrinkage of concrete with SRA is reduced by almost 50% over time

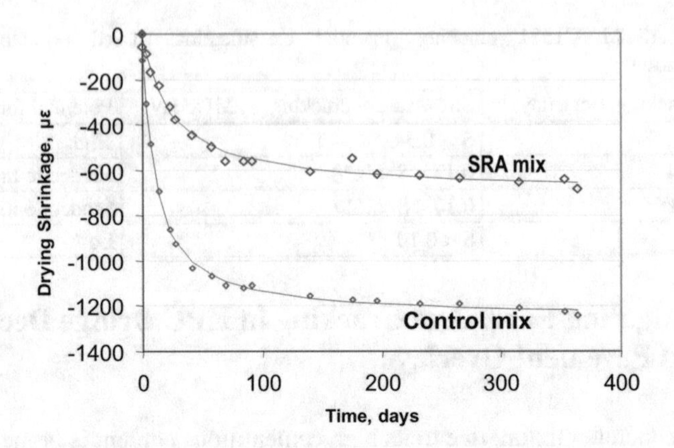

Fig. 8.11 Drying shrinkage of concrete with and without MasterLife SRA 20 admixture. (Courtesy of BASF)

Case Study: Oregon Department of Transportation (ODOT) Research Section Study to Limit Cracking of HPC of Bridge Decks [10]

This study used SRA in concrete mixtures to reduce shrinkage characteristics and to mitigate cracking. The study discusses shrinkage threshold limits for specifications and provides a test procedure that can be followed for evaluating HPC to mitigate cracking.

A summary of the concrete materials, mixture proportions, and drying shrinkage results and other findings is described below.

8.7 Materials

Cementitious Materials: ASTM C150 Type I/II, an ASTM C618 Class F fly ash, and an ASTM C1240 silica fume.

Admixtures: An ASTM C494 Type F polycarboxylate-based high-range water reducer; a SRA (Eclipse 4500), at a dosage rate of 2% of the total cementitious materials by mass. Coarse and Fine Aggregates: siliceous aggregate sources were local river gravel and river sand, and a lightweight aggregate (FLWA) of expanded shale was used as a partial replacement of normal sand. The properties of the aggregates are shown in Table 8.2.

Concrete mixture proportions used in this study are shown in Table 8.3. Fresh and hardened concrete properties for control mix (HPC1) and SRA 1 Mix are shown in Table 8.4.

Table 8.2 Aggregates properties (as received)

Specific gravity		Absorption capacity (%)	Desorption capacity (%)	Fineness modulus
Local sand	2.41	3.08	–	3.0
Local gravel (3/4" MSA)	2.44	2.58	–	7.1
Bend sand	2.54	2.58	–	2.9
End gravel (3/4" MSA)	2.59	2.27	–	7.5
Medford sand	2.48	3.46	–	2.6
Medford gravel (3/4" MSA)	2.53	3.17	–	7.2
Santosh sand	2.58	2.74	–	3.3
Santosh gravel (1" MSA)	2.62	2.04	–	6.7
Limestone (Ontario, Canada)	2.68	0.58	–	6.5
Expanded shale	1.55	17.50	16.0	2.7

Table 8.3 Concrete mixture proportioning

Mixture	Cement (kg/m3)	Fly ash (kg/m3)	Silica fume (kg/m3)	Water (kg/m3)	Coarse aggregate (kg/m3)	Sand (kg/m3)	FLWA (kg/m3)	SRA (kg/m3)
HPC 1	249	112	15	139	1071	659	–	–
SRA	249	112	15	131	1071	659	–	7.5
FLWA	249	112	15	139	1071	400	164	–
SYN	249	112	15	131	1071	400	164	7.5

Figures 8.12 and 8.13 show the shrinkage development curves for different mixtures with 3-day and 14-day wet cure conditions, respectively. By incorporating SRA in HPC control mixture, the shrinkage is significantly reduced for both 3-day and 14-day wet cured samples (SRA1 and SRA 2) versus the control samples without SRA (HPC1 and HPC2).

It is noted in Table 8.5 that SRA mixtures significantly prolonged the time to cracking and decreased the stress rate. SYN mixtures containing expanded shale fine aggregate showed the lowest stress rate and prolonged the time to cracking.

Figure 8.14 shows the relationship between CPI and time to cracking. The figure is divided in four zones based on time to crack. A general trend that follows is that a lower CPI falls into a lower risk zone.

8.8 Recommendations of Various State DOTs on Drying Shrinkage Limits for HPC

Department of Transportation (DOT) in the various states of the USA have been engaged in evaluating alternative materials to control drying shrinkage cracking of concrete bridge decks. Some of the established protocols as summarized below:

Table 8.4 High-performance concrete (SRA 1) with 2%SRA and control mixture (HPC1) test results

	SRA 1 (3-day moist cure)	HPCI control (3-day moist cure)
Designation	Quantities	Quantities
Water/cementitious material ratio	0.37	0.37
Cement type I-II, lb./cy(kg/m^3)	437(249)	437(249)
Silica fume, lb./cy (kg/m^3)	25(15)	25(15)
Fly ash class F, lb./cy(kg/m^3)	189 (112)	189 (112)
Water, lb./cy(kg/m^3)	220(131)	235(139)
Sand, SSD, lb./cy(kg/m^3)	1113(659)	1113(659)
#57 stone, SSD, lb./cy(kg/m^3)	1808(1071)	1808(1071)
Darex AEA, oz./ yd.(ml/m^3)	6.0(232)	6.0(232)
WRDA 35, oz./ yd.(ml/m^3)	22(851)	22(851)
SRA, lb./yd^3 (kg/m^3)	12.65 (7.5)	–
Initial tests after batching (at plant)		
Concrete temperature, F (C)	71(21.6)	71(21.4)
Initial slump, inches (mm)	9 (105)	5 (1250
Air content, %	5.5	6
Unit weight, lb./cf.(kg/m^3)	141 (8.8)	14 (8.99)
Hardened tests		
Compressive strength, 28 days, psi (MPa)	4816 (33.2)	4176 (28.8)
Modulus of elasticity, 28 days, psi (GPa)	4060,000 (28)	3,320,000 (22.9)
Splitting tensile strength, psi(MPa)	576 (3.97)	497 (3.42)
Length change (ASTM C157), at 28 days, %	0.034	0.066
Time to cracking (ASTM C1581), average	17 days	4.9 days
Stress rate (psi/day), average	0.087	0.344

Texas DOT (2006): Use SRA, polypropylene fibers, shrinkage compensating cement, or high-volume fly ash (minimum 50%) to reduce shrinkage cracking.

Virginia DOT (2004): For concrete containing SCM, limit drying shrinkage to 0.04% length change at 28 days and 0.04% at 90 days.

Kansas DOT (2005): Use coarser cement (Type II) and SRA and longer cure time to reduce shrinkage.

New Jersey DOT (2007): Limit free shrinkage to 450 microstrain at 56 days to resist cracking of HPC bridge decks. Use a coarse aggregate to fine aggregate ratio >1.5 to reduce cracking potential.

Washington DOT (2010): Limit free shrinkage to 320 microstrain at 28 days. Use SRA in concrete to reduce free shrinkage and restrained shrinkage cracking. Optimize gradation and reduce cement content.

Ohio DOT (2002–12): Use coarse aggregate with higher absorption capacity (>1.0) to reduce cracking.

California DOT (2011): Use SRA and SCM to reduce drying shrinkage.

Federal HWA (2012): Limit free shrinkage to 300 microstrain at 28 days; 500 microstrain long term.

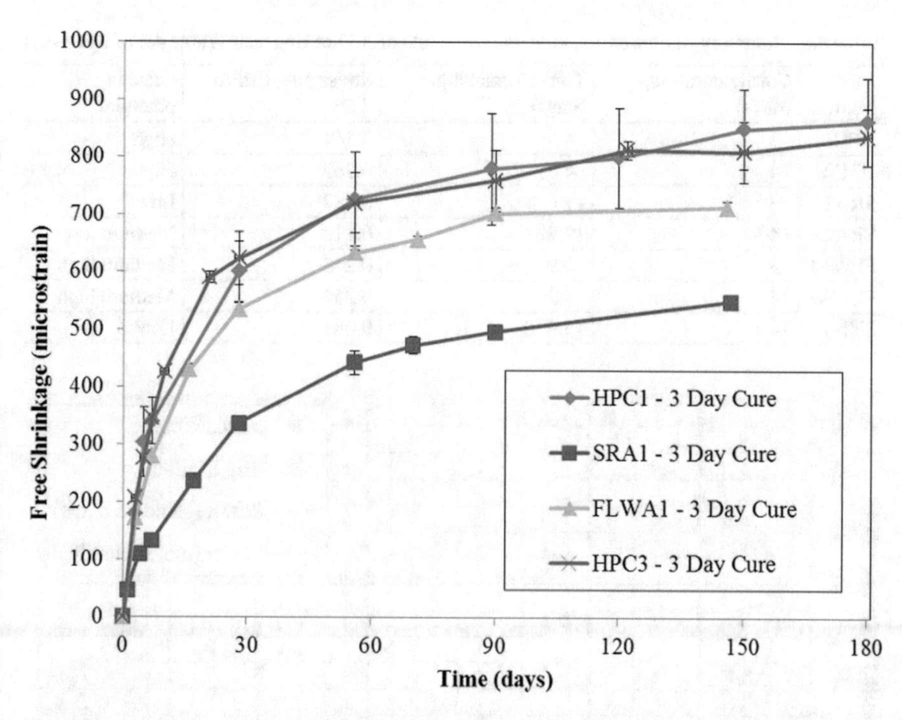

Fig. 8.12 Free shrinkage versus drying time, 3-day cure, effect of shrinkage mitigation methods. (Figure 4.1 of Ref. (9))

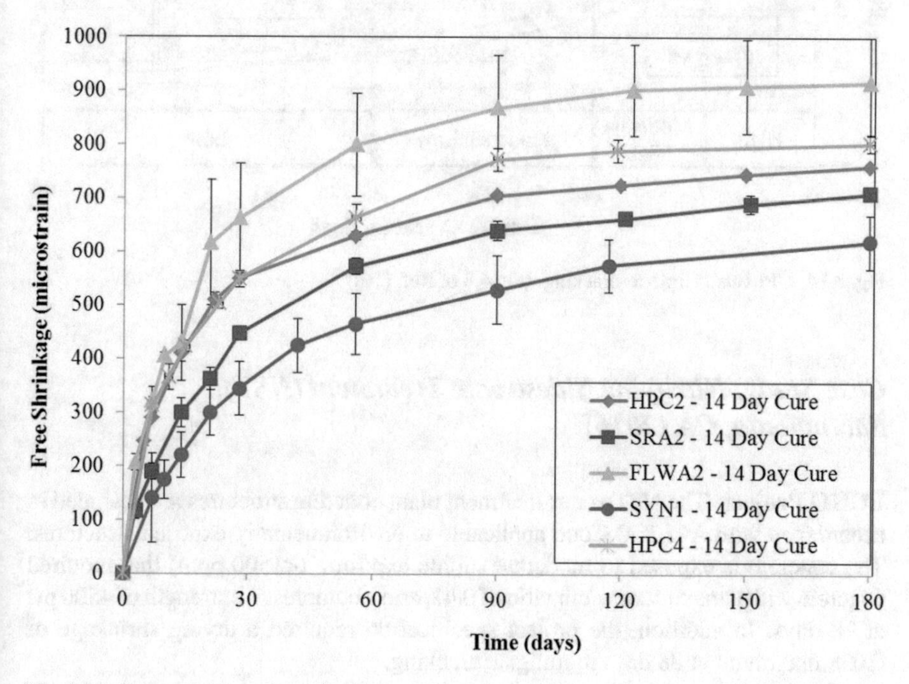

Fig. 8.13 Free shrinkage versus drying time, 14-day cure, effect of shrinkage mitigation methods. (Figure 4.2 of Ref. (10))

Table 8.5 Summary of time to cracking and stress rate of ASTM ring tests (Table 4.5 of Ref. [10])

Mixture	Curing duration, (days)	Time to cracking, (days)	Stress rate, (MPA/ day)	Cracking potential
HPC1	3	4.9	0.344	High
HPC2	14	4.2	0.369	High
SRA1	3	17	0.087	Low
SRA2	14	14.2	0.112	Medium low
FLWA1	3	6.9	0.245	Medium high
FLWA2	14	7.7	0.254	Medium high
SYN1	14	15.9	0.081	Low

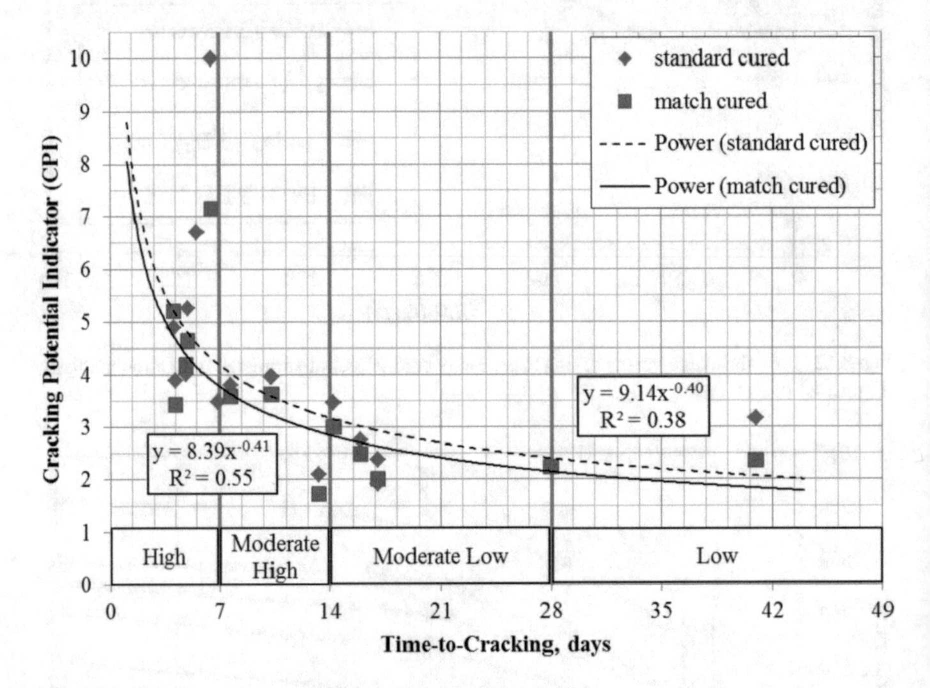

Fig. 8.14 CPI versus time-to-cracking. (Fig 4.8 of Ref. [10])

Case Study: Nitrifying Sidestream Treatment (NST), Sacramento, CA (2016)

ECHO Project The NST water treatment plant concrete structures are designed in accordance with ACI 350 Code applicable to environmentally exposed structures. The concrete is exposed to moderate sulfate exposure (<1500 ppm) that required concrete with a maximum w/cm ratio of 0.42, and a compressive strength of 4500 psi at 28 days. In addition, the project specification required a drying shrinkage of 0.038 maximum at 28 days to mitigate cracking.

The trial mixtures were tested in a testing laboratory in 2016. The aggregates were nonreactive, sound, and low absorption. The coarse aggregate was #57 stone (1-inch), and sand conformed to ASTM C33.

Cement was Type II/V, furnished by Nevada Cement Company, conformed to ASTM C150. It had low alkali (0.56%), with C_3A of 4% and 2(C_3A + C4AF) of 19%, C_3S + 4.75*C_3A of 77%, and Blaine fineness of 4120 cm^2/kg. Fly ash, furnished by Nevada Cement Company, conformed to ASTM C618. It had low alkali (1.4%) and low SO3 (0.2%).

The selected concrete proportions ECH038P are shown in Fig. 8.15. The actual drying shrinkage, measured in accordance with ASTM C157, with 4 in. × 4 in. × 11.25 in. bars, after initial 7-day moist cure, is shown in Table 8.6.

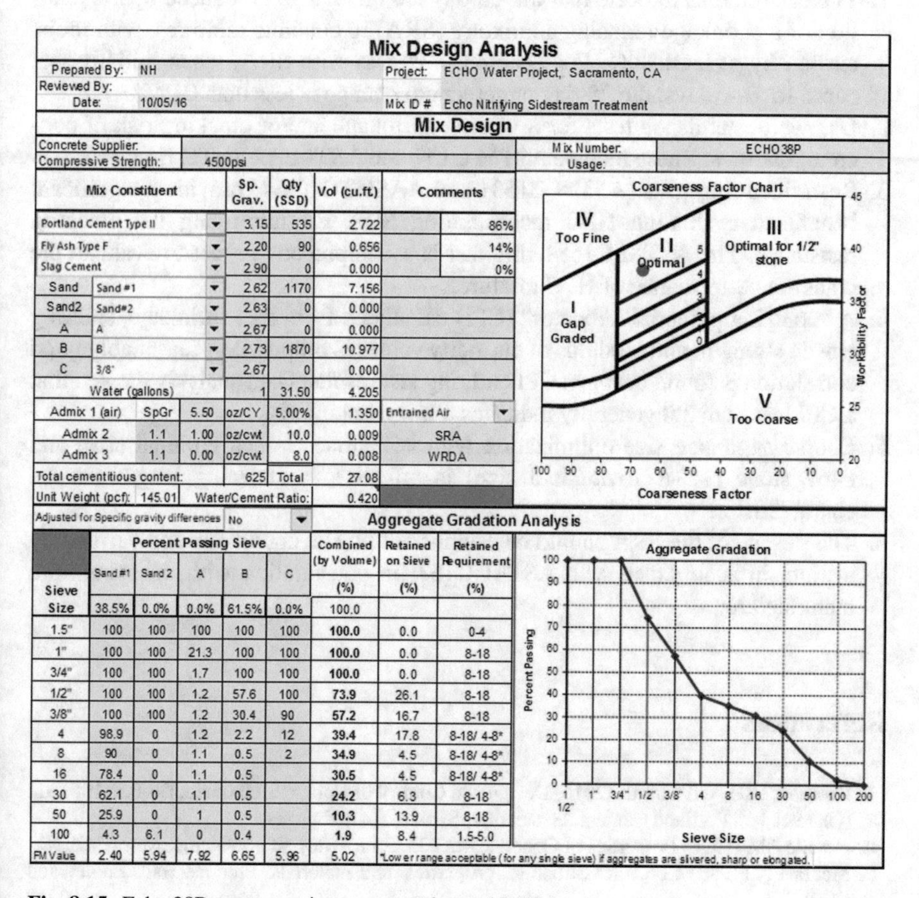

Mix Design Analysis

Prepared By: NH — Project: ECHO Water Project, Sacramento, CA

Reviewed By:

Date: 10/05/16 — Mix ID # Echo Nitrifying Sidestream Treatment

Mix Design

Concrete Supplier: — Mix Number: ECHO38P

Compressive Strength: 4500psi — Usage:

Mix Constituent			Sp. Grav.	Qty (SSD)	Vol (cu.ft.)	Comments
Portland Cement Type II			3.15	535	2.722	86%
Fly Ash Type F			2.20	90	0.656	14%
Slag Cement			2.90	0	0.000	0%
Sand	Sand #1		2.62	1170	7.156	
Sand2	Sand#2		2.63	0	0.000	
A	A		2.67	0	0.000	
B	B		2.73	1870	10.977	
C	3/8"		2.67	0	0.000	
Water (gallons)			1	31.50	4.205	
Admix 1 (air)	SpGr	5.50	oz/CY	5.00%	1.350	Entrained Air
Admix 2	1.1	1.00	oz/cwt	10.0	0.009	SRA
Admix 3	1.1	0.00	oz/cwt	8.0	0.008	WRDA
Total cementitious content:			625	Total	27.08	
Unit Weight (pcf): 145.01		Water/Cement Ratio:			0.420	
Adjusted for Specific gravity differences	No					

Aggregate Gradation Analysis

Sieve Size	Sand #1	Sand 2	A	B	C	Combined (by Volume) (%)	Retained on Sieve (%)	Retained Requirement (%)
	38.5%	0.0%	0.0%	61.5%	0.0%	100.0		
1.5"	100	100	100	100	100	100.0	0.0	0-4
1"	100	100	21.3	100	100	100.0	0.0	8-18
3/4"	100	100	1.7	100	100	100.0	0.0	8-18
1/2"	100	100	1.2	57.6	100	73.9	26.1	8-18
3/8"	100	100	1.2	30.4	90	57.2	16.7	8-18
4	98.9	0	1.2	2.2	12	39.4	17.8	8-18/ 4-8*
8	90	0	1.1	0.5	2	34.9	4.5	8-18/ 4-8*
16	78.4	0	1.1	0.5		30.5	4.5	8-18/ 4-8*
30	62.1	0	1.1	0.5		24.2	6.3	8-18
50	25.9	0	1	0.5		10.3	13.9	8-18
100	4.3	6.1	0	0.4		1.9	8.4	1.5-5.0
FM Value	2.40	5.94	7.92	6.65	5.96	5.02		

*Lower range acceptable (for any single sieve) if aggregates are slivered, sharp or elongated.

Fig. 8.15 Echo 38P concrete mixture proportions with SRA

Table 8.6 ASTM C157 length change for ECHO30P concrete

Age	Average length change, %
7 days	0.012
14 days	0.02
21 days	0.028
28 days	0.032

8.9 Conclusions and Recommendations

Based on the shrinkage results and referenced studies, the following recommendations were made:

1. The case studies indicate that the 28-day length change is reduced by the addition of a shrinkage reducing admixture (SRA) to concrete mixtures, with moist curing, by at least 30%. For concrete mixtures with SRA, when initially wet cured for 3–7 days, the 28-day target length change is less than 0.04%.
2. Use well-established tests for assessing shrinkage and/or cracking risk of concrete mixtures. These include ASTM C157 and ASTM/AASHTO ring tests.
3. Restrained ring tests (ASTM C1581 and AASHTO T334) provide a significant benefit in evaluating HPC incorporating SRA for improving the cracking resistance. The ASTM C1581 ring test is a comprehensive test to evaluate the cracking performance of HPC mixtures.
4. A "cracking potential indicator" (CPI) calculated from free shrinkage, splitting tensile strength, and modulus of elasticity values is proposed. A reasonably good correlation is found between CPI and ring test results. Data analysis showed that a CPI less than 3.0 generally indicates a low cracking risk.
5. Coarse aggregate size optimization with #57 stone (1-inch nominal size) and #467 stone (1.5-inch nominal size) is effective in reducing length change characteristics.
6. The dosage of the SRA should be established by site conditions and verified by testing, in accordance with ASTM C157, in consultation with the admixture manufacturer.

References

1. Kosmatka SH, Wilson ML (2016) Design and Control of Concrete Mixtures, EB 001.16th edn (Chapter 13). Portland Cement Association, Skokie, IL. 632 pages
2. Neville AM (1996) Properties of Concrete.4th edn. John Wiley and Sons, Inc, New York, NY
3. Mehta PK (1986) Concrete: Structure, properties, and materials. Prentice-Hall, Englewood Cliffs, NJ
4. Long term Study of Cement Performance in Concrete (1947) Test of 28 Cements of Greemn Mountain Data, Materials Laboratories Report No. C-345, US. Department of the Interior, Bureau of Reclamation, Denver

5. Jackson FH (1955) Long term study of cement performance in concrete- Chapter 9. In: Correlation of the results of laboratory tests with field performance under natural freezing and thawing conditions. Research Department Bulletin X060, Portland Cement Association
6. Gebler S, Klieger P (1986) Effect of Fly ash and some of the physical Properties of concrete. , Research and Development Bulletin RD 089, Portland Cement Association
7. Aitcin PC (1999) Does concrete shrink or does it swell. Concrete International, American Concrete Institutue, MI, pp 77–80
8. Wan B, Foley CM, Komp J (2010) Concrete cracking in new bridge decks and overlays, Wisconsin highway research program report. WHRP 10-05:2010
9. Whiting DA, Detwiler RJ, Lagergren ES (2000) Cracking tendency and drying shrinkage of silica fume concrete for bridge deck applications. ACI Mater J 2000:71–78
10. Ideker JH, Fu T, Deboodt T (2013) Development of shrinkage limits and testing protocols for ODOT HPC. Oregon Department of Transportation, Research Section, Report SPR-728, FHWA-OD-RD-14-09

Chapter 9
Quality Control During Production

9.1 Quality Assurance (QA)

The QA program should conform to the applicable requirements of the specifications and supplemented by the specified codes and standards. The QA program should be approved prior to start of construction. A typical QA document submittal requirements schedule is listed in Table 9.1.

The QA should, as a minimum, address the following:

1. Concrete supplier responsibilities
2. On-site batch plan's responsibilities
3. Non-conformances
4. Audits and hold points
5. Shipping, packaging, and transportation
6. Storage requirements
7. QA documentation

9.2 Quality Control (QC) Plan

The QC plan should identify personnel, procedures, instructions, tests, records, and forms to be used. The contractor may submit their standard QC plans for fabrication and erection if they meet the following requirements.

The QC plan should include, as a minimum, the following to cover all fabrication, preassembly, and erection operations, both on-site and off-site, including work by subcontractors, fabricators, suppliers, and purchasing agents:

A. A description of the QC organization, including a chart showing the lines of authority. The staff should include a QC system manager who should report to the project manager or someone higher in the contractor's organization.

N. Hasan, *Durability and Sustainability of Concrete*,
https://doi.org/10.1007/978-3-030-51573-7_9

Table 9.1 Typical document submittal requirements

Number	Document	Delivery requirements	Submit for
1	Quality assurance program manual	With bid	Approval
2	Concrete production plan and procedures	Upon award	Approval
3	Shipping, packaging, and transportation procedures	Upon award	Information
4	Storage requirements	With material shipment	Information
5	Control of materials	With material shipment	Information
6	Cements material certification	With material shipment	Information
7	Supplementary material certification	With material shipment	Information
8	Admixtures certification	With material shipment	Information
9	Aggregate certification	With material shipment	Information
10	NRMCA certification	Prior to production of concrete	Information
11	Record of equipment calibration	Prior to production of concrete	Information
12	Record of batching tolerances	Prior to production of concrete	Information
13	Trial batch plan for concrete mixtures	Prior to trial mixture development	Approval
14	Concrete batching records	As requested	Information
15	Non-conformances report	Upon occurrence	Approval
16	Audit reports	Upon request	Information
17	Concrete test reports	With production	Information

B. The name, qualifications, duties, responsibilities, and authorities of each person assigned a QC function.

C. A copy of the letter of authority of the contractor's quality control program manager signed by an authorized official of the contractor and which describes the responsibilities and delegate's sufficient authorities to adequately perform the functions of the QC program manager including authority to reject materials and work and to stop work which is not in compliance with the contract.

D. Procedures for scheduling, reviewing, certifying, and managing submittals, including those of the contractor, subcontractors, off-site fabricators, suppliers, and purchasing agents. Control, verification, and acceptance testing procedures for each specific test to include the test name, section paragraph requiring test, feature of work to be tested, test frequency, and person responsible for each test.

E. Procedures for tracking preparatory, initial, and follow-up control phases and control, verification, and acceptance tests including documentation.

F. Procedures for tracking deficiencies from identification through acceptable corrective action. These procedures will establish verification that identified deficiencies have been corrected.

G. Reporting procedures, including proposed reporting formats and samples.

A list of the definable features of work. A definable feature of work is a task which is separate and distinct from other tasks and has separate control requirements.

ACI 301 Requirements For buildings and infrastructure concrete, ACI 301 provides guidance on testing of concrete during construction. For acceptance purposes, composite sample for each 100 cubic yards of each concrete mixture placed is taken from the end of truck discharge. Concrete cylinders are molded in accordance with ASTM C31 and tested for compressive strength at 7 and 28 days of age. ACI 301 requires that a minimum of one specimen is tested at 7 days and a minimum of two specimens are tested for strength (ASTM C39), after curing under laboratory conditions. The compressive strength for acceptance is based on the average strength of two cylinders tested at 28 days. In addition to strength test, ACI 301 requires tests for slump (ASTM C143), air content (ASTM C173, C231, or C138), and temperature (ASTM C1064) for each composite sample taken from the truck. ACI 301 provides the option for additional concrete testing during construction, as required by the architect/engineer or directed in the contract documents.

ACI 359 Requirements For nuclear facilities, the ASME NQA-1-2008 (Ref. [1]) provides guidance on testing of concrete during construction, which requires that the in-process tests and testing frequency for concrete shall be in accordance with ACI 359 (Ref. [2]) and ASME Boiler and Pressure Vessel Code, Section III, Division 2.

Quality Control Tests Field quality control requirements for concrete are established in the contract documents. Quality control should focus on the critical attributes of the concrete materials and the final product that influences mixture consistency and compressive strength. The specifier should specify the extent of quality control during production.

Commercial-Grade Dedication Plans (CGDP) Commercial-grade dedication is a process by which a *commercial-grade item (CGI)* is designated for use as a basic component. For nuclear facilities, this *acceptance process* is undertaken to provide reasonable assurance that a CGI to be used as a basic component will perform its intended safety function and, in this respect, is deemed equivalent to an item designed and manufactured under a 10 CFR Part 50, *Appendix B, quality assurance program* for nuclear plants.

Commercial-grade dedication plans (CGDP) for concrete materials including cement, fine and coarse aggregates, air entraining and chemical admixtures, and water should be prepared. The material sources, technical evaluation, acceptance test methods, and critical characteristics for each CGDP should be identified, in accordance with the project specifications, tested and approved prior to use.

Responsibility for Quality Control The contractor is responsible for quality control and to establish and maintain an effective quality control program. The quality control program shall consist of plans, procedures, and organization necessary to produce an end product, which complies with the contract requirements. The program shall cover all fabrication, pre-assembly, and erection operations, both on-site and off-site, and shall be keyed to the proposed construction sequence. The contractor's project manager will be held responsible for the quality of work.

Acceptance of Plan Acceptance by the owner of the QC plan is required prior to the start of fabrication. Acceptance is conditional and will be predicated on satisfactory performance during the work. The owner reserves the right to require the contractor to make changes in the QC plan and operations, as necessary, to obtain conformance with contract requirements.

9.3 QC Control Compliance Testing

Daily checks should be performed on the ongoing work including control testing to assure continued compliance with contract requirements. The checks should be made a matter of record in the QC documentation.

Testing Procedure Tests that are specified or required should be performed to verify that control measures are adequate to provide a product which conforms to contract requirements. Testing includes operation and/or acceptance tests when specified. The services of a owner-approved independent testing laboratory should be enlisted. A list of tests to be performed a part of the QC plan is included in Table 9.2. The following activities should be performed and recorded and the following data provided:

A. Verify that testing procedures comply with contract requirements.
B. Verify that facilities and testing equipment are available and comply with testing standards.
C. Check test instrument calibration data against certified standards.
D. Verify that recording forms and test identification control number system, including all of the test documentation requirements, have been prepared.
E. Results of all tests taken, both passing and failing tests, will be recorded on the QC report for the date taken. Section paragraph reference, location where tests were taken, and the sequential control number identifying the test will be given.

Table 9.2 In-process testing of concrete during production

Description	Test method	Testing frequency
Fine aggregate grading	ASTM C33	Daily during production
Fine aggregate grading	ASTM C33	Daily during production
Sampling concrete	ASTM C172	Every 50 cy or thereof
Air content	ASTM C231	Every 50 cy or thereof
Slump	ASTM C143	Every 50 cy or thereof
Slump flow	ASTM C1611	Every 50 cy or thereof
Unit weight	ASTM 138	Every 50 cy or thereof
Bleed water	ASTM C232	As directed
Temperature	ASTM C1064	Every 50 cy or thereof
Set time	ASTM C403	As directed
Making strength specimens	ASTM C31	Every 100 cubic yards
Compressive strength tests	ASTM C39	Every 100 cubic yards

Actual test reports may be submitted later, if approved, with a reference to the test number and date taken. An information copy of tests performed by an off-site or commercial test facility will be provided directly to the owner.

Documentation Records of QC operations, activities, and tests performed shall be maintained at the site, including the work of subcontractors and suppliers. These records should include factual evidence that required quality control activities and/or tests have been performed including but not limited to the following:

A. Contractor/subcontractor and their area of responsibility
B. Test and/or control activities performed with results and references to sections/plan requirements. List deficiencies noted along with corrective action
C. Quality of material received at the site with statement as to its acceptability, storage, and reference to section drawing requirements
D. Submittals reviewed, with contract reference, by whom, and action taken
E. Off-site surveillance activities, including actions taken
F. Job safety and environmental protection evaluations

Submittals of all quality control tests and reports should be as specified in the QC documents. The contractor's QC program manager should be responsible for certifying that all QC submittals comply.

Notification of Noncompliance For any detected noncompliance with the foregoing requirements, the contractor shall take immediate corrective action.

Field Quality Control Field quality control of concrete during production is performed by an independent testing agency, engaged by the owner, engineer, or construction manager, who should be experienced in sampling and testing of concrete in accordance with ASTM E 329.

The testing agency inspects, samples, and tests materials and production of concrete to assure compliance with these specifications. The contractor provides copies of all tests and inspection reports to the owner's representative in a timely fashion.

Batch Plant Before concrete production is started, the contractor provides access for obtaining samples of cement, aggregates, and concrete as required. These facilities are provided at the batching and mixing plant.

Acceptance of Concrete The average of three consecutive strength tests shall be equal to or greater than the specified 28- or 56-day compressive strength. No individual strength shall be more than 500 psi below the specified 28-day compressive strength. Table 9.2 provides in-process testing of concrete materials and concrete during production.

Construction Tolerances Construction tolerances for concrete structures should be established in accordance with ACI 117. Some selected construction tolerances are listed in Tables 9.3 and 9.4.

Table 9.3 Tolerances for cast-in-place reinforced concrete

Item	Description	Tolerance
Vertical alignment	For heights 100 ft. or less	
	Lines, surfaces, and arises	I inch
	Exposed columns and control joints, openings	1/2 inch
Lateral alignments		
	Members	I inch
	In slabs, edges, and openings location	1/2 inch
Level alignment		
	Elevation top of slab	3/4 inch
	Lintels, parapets, and exposed lines	1/2 inch
Cross-sectional dimensions		
	Members, walls, slabs	
	12 inches or less	+ 3/8. −1/4 inch
	>12 inches	+1/2, −3/8 inch
Relative alignment		
	Stairs-difference in height	1/8

Table 9.4 Tolerances for finished formed surfaces

Vertical alignment		
	Surface slope exposed to view	1/4 inch in 10 feet
Abrupt variation		
	Perpendicular to flow	0 to 1/8 inch
	Parallel to flow	1/8 inch
Gradual variation		
	Surface finish (measured by a 5-ft straight edge)	1/8 inch to 1/4 inch
Floor flatness		
	Bull-floated	1/2 inch in 10 feet
	Straight-edged	5/16 inch in 10 feet
	Float finish	3/16 inch in 10 feet
	Trowel finish	3/16inch in 10 feet

9.4 Case Studies: San Roque Project Quality Assurance/ Quality Control (QA/QC)

To meet the challenge of producing high-quality concrete in accordance with the specified requirements for consistency, workability, and strength, an extensive concrete QA/QC program was conducted at site. The QA/QC program included the following:

1. Conducted concrete trial mix designs for various classes of concrete and verify workability, pumpability, initial and final setting times, and compressive strength

2. Obtained daily samples of aggregates from the processing plant conveyor belts and stockpiles and performed gradation tests on coarse and fine aggregates
3. Determined daily moisture content of concrete aggregate stockpile prior to batching
4. Sampled and tested cement and fly ash for compliance with the specification compliance requirement
5. Tested samples of concrete for slump, unit weight, temperature, yield, and compressive strength every 100 cubic meters
6. Inspected and monitored concrete, during and after placement
7. Inspected concrete plant, including scales and batch weights, every 3 months and recalibrate, if required
8. Tested the 28-day and 56-day compressive strengths including for compliance with the ACI 318 and ACI 214 requirements

9.5 Qualification of Concrete Materials

- Cement: Cement conformed to ASTM C-150. Type II cement was used as per project specification.
- Fly ash: Class F fly ash, used as replacement of cement, conformed to ASTM C-618.
- Coarse Aggregates: Gravel or coarse aggregates crushed or rounded were tested for compliance to ASTM C-33. Tests included (1) sieve analysis of coarse aggregates (ASTM C-136), (2) materials finer than No. 200 sieve (ASTM C-117), (3) abrasion loss (ASTM C-131), (4) soundness test (ASTM C-88), (5) clay lumps and friable particles (ASTM C-142), (6) specific gravity (ASTM C-127), (7) unit weights (ASTM C-29), (8) flat and elongated particles, and (9) potential reactivity with alkalis (ASTM C-289).
- Fine Aggregates: Fine aggregates (sand) conformed to the requirements of ASTM C-33.Tests included sieve analysis of fine aggregates (ASTM C-136), (2) materials finer than No. 200 sieve (ASTM C-117), (3) soundness test (ASTM C-88), (4) specific gravity and absorption (ASTM C-128), (5) sand equivalent, (6) unit weights (ASTM C-29), (7) organic impurities in fine aggregates (ASTM C-40), (8) clay lumps and friable particles in aggregates (ASTM C-142), and (9) light weight material (ASTM C-123).
- Water: Water used was clean and free from any deleterious materials or organic impurities and conformed to ASTM C-94.
- Admixtures: Chemical admixtures conformed to ASTM C-494, Types A through G, used depending on formulations, and their purposes for use in concrete included Type A (water reducing), Type B (retarding), Type C (accelerating), Type D (water reducing and retarding), Type E (water reducing and accelerating), Type F (water reducing, high range), and Type G (water reducing, high range, and retarding and air entraining admixture).

ACI 211.1 Mixture Design Factors considered were (1) maximum water-cement or water-cementitious material ratio, (2) minimum cement content, (3) air content, (4) slump, (5) maximum size of aggregates, (6) strength, (7) specific gravity of materials, (8) estimation of mixing water and air content (includes selection of admixture), (9) selection of water-cement or water-cementitious materials ratio, (10) calculation of cement content, (11) estimation of coarse aggregate content, (12) estimation of fine aggregates content, (13) adjustment for aggregate moisture, (14) trial batch adjustments, and (15) trial mix for use in concrete.

Trial mixes were run to supplement and verify the behavior during actual placement situations. Tests conducted with one admixture (HPC) were determined to be satisfactory. Alternate mix designs were developed substituting admixtures PSPR and PDA-25R with HPC to handle this specific problem.

Highlights of the San Roque QC Program (2000–2002):

1. Concrete laboratory services for testing concrete materials and concrete, prestressed/precast.
2. Sampling of cement, fly ash, concrete, and rebars for testing by an independent laboratory in order to verify physical and chemical properties of cementitious materials and concrete placements for conformance with ASTM standards and technical specification.
3. Calibrations of Johnson Ross and Ross batch plants every 6 months.
4. Compressive strength testing machines calibration.
5. Facilitated initial curing of concrete samples at the place of concrete placements by providing coolers and storage facility at Spillway.
6. A comprehensive in-house inspection program was performed to provide round-the-clock inspection services by trained and professional personnel.

Non-conformances and Construction Issues Some typical non-conformances were:

1. Modifying mix design, such as requiring "higher slumps" and "early high strengths" to satisfy field workability conditions.
2. Variations in moisture content of fine sand resulting in inaccurate water quantities affecting ultimate concrete strengths.
3. Deviations in percent passing # 200 sieves above the specified limits for coarse aggregate.
4. Deviations of chemical and physical properties of cementitious materials (i.e., cement, fly ash) causing variations in strength characteristics.
5. Pumping equipment limitations – 52 meters pump cement pump for mass concrete mix with 3″ aggregate – requiring telebelt/ bucket to place mass concrete (M113A).
6. Spillway chute and ogee placements challenges due to steep topography (27% slope) and F4 finish requirements. Several changes were made in placement methodology as well as modifications in mix design (balancing of aggregate proportions) to improve workability in order to achieve desired results. After initial difficulties, however, the finishing of chute slab was acceptable.

Compressive Strength Results The concrete compressive strength exceeded the specified strength often as high as twice the required strength at 56 days. Some of the factors contributing to the high strength include holding of allowable water, high cement content, high Blaine value (fineness of cement grind) of cement, etc.

Figure 9.1 and Table 9.5 show Spillway Concrete Mixture AAA – 2 M% compressive strength results at 7 and 56 days during production.

Spillway Chute Surface Flatness Quality Control The specified surface flatness tolerance for the spillway chute is more stringent than the accepted industry tolerance (Ref. [3]).

Surface irregularity	Tolerance	Ref. [3] tolerance (40 ft/sec)
Abrupt irregularity (offset	0 mm	3 mm
Gradual irregularity (slope)	1.25 mm	6 mm
Gradual irregularity (offset	6 mm	6 mm

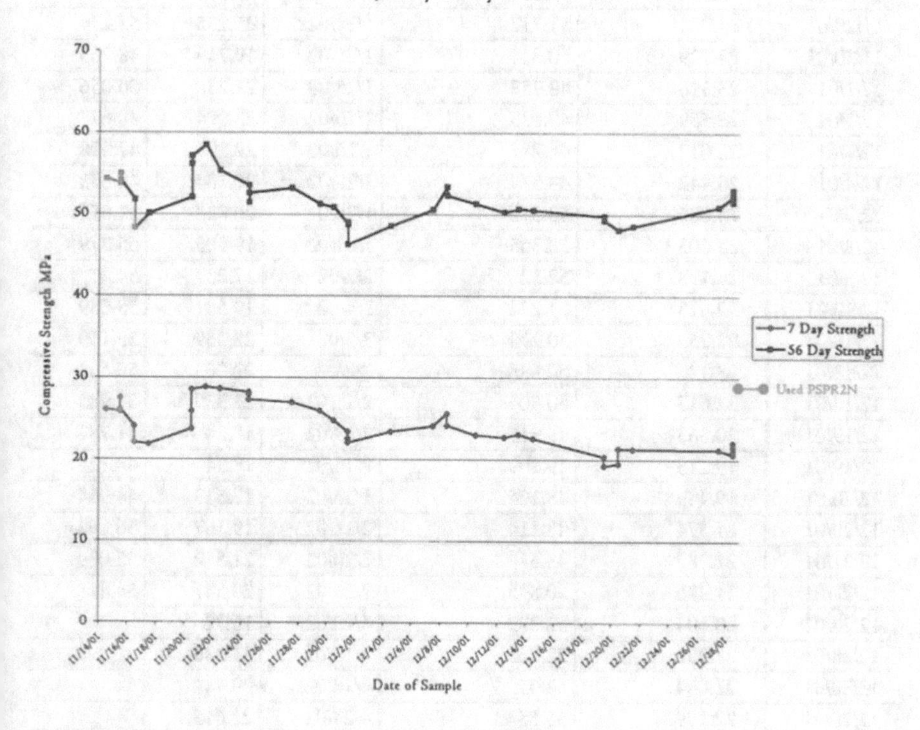

Fig. 9.1 Spillway concrete 7-day and 56-day strength results (MPa) (11/14/01–5/1/02 period)

Table 9.5 Spillway concrete mixture AAA2 7-day and 56-day strength moving averages (11/14/01–5/1/02 period)

Date of sample	Moving ave. 3 7-day strength	Moving ave. 3 56-day strength	Date of sample	7-day strength	56-day strength
11/14/01			1/3/02	15.858	46.746
11/04/01			1/3/02	18.547	50.056
11/14/01	26.039	54.283	1/4/02	23.58	52.883
11/15/01	25.609	53.233	1/4/02	23.442	51.676
11/15/01	27.454	54.882	1/4/02	21.788	50.47
11/15/01	26.085	54.101	1/7/02	24.683	58.537
11/16/01	24.04	51.734	1/8/02	20.96	54.4
11/16/01	22.018	48.263	1/9/02	22.339	51.918
11/17/01	21.834	50.044	1/9/02	22.201	51.4
11/20/01	23.707	52.055	1/10/02	21.925	50.021
11/20/01	25.855	56.123	1/14/02	18.099	43.747
11/20/01	28.464	57.077	1/16/02	24.959	54.4
11/21/01	28.774	58.445	1/18/02	22.891	49.642
11/22/01	28.556	55.284	1/22/02	19.582	55.779
11/24/01	28.154	53.457	1/23/02	24.684	55.848
11/24/01	27.901	51.435	1/24/02	27.821	53.986
11/24/01	27.257	52.573	1/24/02	25.097	53.055
11/27/01	26.935	53.17	1/24/02	25.511	57.295
11/29/01	25.924	51.217	1/24/02	25.235	53.294
11/30/01	24.729	50.814	1/25/02	19.719	48.263
12/1/01	23.396	48.757	1/25/02	21.236	50.056
12/1/01	22.569	49.033	1/26/02	23.994	48.677
12/1/01	22.017	46.287	1/26/02	22.891	47.298
12/4/01	23.442	48.574	1/28/02	20.684	49.642
12/7/01	24.086	50.573	1/29/02	20.822	51.952
12/8/01	25.603	53.365	1/31/02	19.305	55.779
12/8/01	24.178	52.32	2/2/02	17.237	54.4
12/10/01	23.028	51.274	2/4/02	19.581	55.089
12/12/01	22.753	50.274	2/5/02	22.339	53.159
12/13/01	23.12	50.688	2/8/02	22.063	55.503
12/14/01	22.615	50.504	2/11/02	22.339	57.985
12/19/01	20.363	49.815	2/13/02	18.34	41.782
12/19/01	19.213	49.309	2/14/02	18.34	48.125
12/20/01	19.489	48.125	2/15/02	17.513	44.954
12/20/01	21.374	48.114	2/18/02	19.167	56.192
12/21/01	21.282	48.574	2/20/02	21.512	55.089
12/27/01	21.236	50.895	2/23/02	20.547	54.4
12/28/01	20.684	52.239	4/10/02	16.72	
12/28/01	21.052	52.963	4/13/02	21.236	
12/28/01	21.604	52.055	4/18/02	29.441	
12/28/01	22.109	51.55	4/24/02	22.615	
			4/29/02	19.995	
			5/1/02	21.65	

Both abrupt and gradual surface offsets were surveyed with a 3 m straight edge template with 50-mm high chairs. Measurements were taken at 1 m intervals in both (transverse and longitudinal) directions of all the bays in Chutes 1 through 6. At few locations where the surface flatness tolerances were exceeded, the specification called for concrete removal by grinding in accordance with Ref (3).

The results indicate that for Chute 1, immediately downstream of the gate sill, where flow velocity range from 46 to 75 feet (14 to 23 meters) per second and the flow is not aerated, all the measurements were within the 1 in 20 bevel limit, required by Ref. [3]. For Chutes 2 through 6, where the flow velocity range from 78 to 111 feet per second and the flow is fully aerated, only two measurements exceeded the 1 in 50 bevel limit, required by Ref. [4]. Since the flow is fully aerated in Chutes 2–6, the as-built flatness meets the Ref. [4] bevel requirements (1:8). As a result, no grinding of the chute surface was required.

Case Study: Detroit Edison Fermi 2 Station Duct Bank

The construction of the Electric Duct Bank at the Detroit Edison Fermi 2 Nuclear Power Station was performed during 2009–2011 period. The testing of concrete materials was performed in accordance with nuclear industry standards as described herein (Table 9.6).

Concrete Aggregates Aggregates conformed to ASTM C33, Standard Specification for Concrete Aggregates, and the following requirements:

Coarse aggregate (number 57 or 1-inch nominal maximum size (Stoneco Denniston Pit)):

LA abrasion (ASTM C131) weight loss: 45% at 500 revolutions
Alkali-silica reactivity expansion (ASTM C1260/C1567): 0.1% at 16 days
Material passing No. 200 sieve (ASTM C117): 2% maximum

Fine aggregate (CTE sand and gravel pit):

Alkali-silica reactivity expansion (ASTM C1260/C1567): 0.1% at 16 days
Limit for clay lumps and friable particles: 8%
Percent passing # 50 sieve: 25–40% (provided the fineness modulus is within 0.20 of the average fineness modulus)

Since the local fine and coarse aggregate materials grading, including fines, were generally different than permitted by the ASTM C33 specification, the grading range, percent finer than 200 sieve, and friable particles limits were relaxed for the Fermi Electrical Duct Bank concrete materials. Specifically the fine aggregate percentage passing the No. 50 sieve were relaxed to 5–40% (ASTM C33 range: 5–30%) percent, provided the fineness modulus (FM) of the fine aggregate was within 0.20 of the average FM of the previous ten tests.

Table 9.6 ACI 349, ACI 359, and ANSI N45.2.5 testing requirements

Material/test	Industry standards (see note 2)		
	ACI 349	ACI 359/ NQA-1	ANSI N45.2.5
Cement			
1. Chemical composition	Mfgr's CMTR with each shipment	Each 1,200 tons	Each 1,200 tons
2. Loss on ignition		Not required	
3. Insoluble residue		Not required	
4. Normal consistency		Not required	
5. Air content of mortar		Not required	
6. Fineness		Each 1,200 tons	
7. Autoclave expansion		Each 1,200 tons	
8. Compressive strength		Each 1,200 tons	
9. Vicat initial set time		Each 1,200 tons	
Silica fume			
SiO_2	Mfgr's CMTR with initial shipment	Not mentioned	Not mentioned
Moisture content			
Loss on ignition			
Oversize			
Strength activity index			
Specific surface			
Fly ash			
$SiO_2 + Al_2O_3 + Fe_2O_3$	Mfr's CMTR with initial shipment	Each 2,000 tons	Each 200 tons
SO_3		Not required	
Moisture content		Not required	
Loss on ignition		Each 400 tons	
Fineness		Each 400 tons	
Strength activity index		Not required	
Autoclave expansion		Each 2,000 tons	
Air-entraining agent			
Infrared analysis	Each shipment	Each shipment	Each shipment
pH	Mfr's C of C		Not required
Residue by oven drying	Mfr's C of C		Not required
Water-reducing agents			
Infrared analysis	Each shipment	Each shipment	Each shipment
Specific gravity	Mfr's C of C	Each shipment	Not required
Residue by oven drying	Mfr's C of C	Each shipment	Not required
Chloride content	Mfr's C of C	Not required	Not required
Fine aggregate			
Gradation	Daily	Each 2,000 yds	Daily
Finer than #200 sieve	Daily	Each 2,000 yds	Daily
SG & absorption	Not required	Semiannual	Not required
Organic Impurities	Initially	Each 2,000 yds	Weekly

(continued)

Table 9.6 (continued)

Alkali reactivity	Initially	Not required	Semiannual
Petrographic analysis	Not mentioned	Semiannual	Not required
Coarse aggregate			
Gradation	Daily	Each 2,000 yds	Daily
Finer than #200 sieve	Daily	Each 2,000 yds	Daily
SG & absorption	Not required	Semiannual	Not required
LA abrasion	Initially	Semiannual	Semiannual
Alkali reactivity	Initially	Not required	Semiannual
Petrographic analysis	Not mentioned	Semiannual	Not required

Notes (1) Manufacturer's CMTR with each shipment and designated tests with initial and every other shipment. *(2)* ACI 359 applies to reactor containment and ACI 349 applies to all other power structures. ANSI N45.2 was adopted by NRC Regulatory Guide 1.94. ANSI N45.2.5 has been superseded by NQA-1 which currently references ACI 359 concrete material testing frequencies. *(3)* Test is only required when aggregate sources change or there is a significant change in annually required test results

Based on the qualification tests on fine aggregates (sand), the base fineness modulus (FM) for fine aggregate was 2.65. Since a 0.20 FM variation is allowed by the specification, the acceptable FM range, during production, was 2.45 and 2.85 (Fig. 9.2).

Cement The cement (LaFarge Plant Detroit, MI) conforms to an ASTM C150 Type I/II.

Fly Ash Fly ash (Holcim, Dundee, MI) conforms to ASTM C618 Class C. Fly ash in amounts of approximately 20% by weight of cement replacement was used to mitigate sulfate attack and to improve durability of the concrete.

Silica Fume Silica fume (LaFarge Plant) conforms to ASTM C1240. Silica fume in amounts of approximately 10% by weight of cement replacement was used to improve durability of concrete.

Admixtures Air-entraining admixture (GRT Polychem AE), conforming to ASTM C260, was used in all concrete. The air content requirements for concrete were 5–7.5%. GRT KB-1000, a water-reducing admixture, conforming to ASTM C494 Type D, and GRT Melchem Superplasticizer, conforming to ASTM C94 Type F, were added at the batch plant to each concrete batch for workability. The slump requirements were 5–8 inches.

Concrete Design Mixture The concrete design mix proportions for class (minimum design compressive strength 4000 psi at 28 days) were established in accordance with the project specification such that the average compressive strength of the design mix was 5200 psi or more at 28 days.

The aggregates sources were qualified for use after the materials were tested for petrographic analysis and alkali reactivity tests. Fly ash and silica fume were added

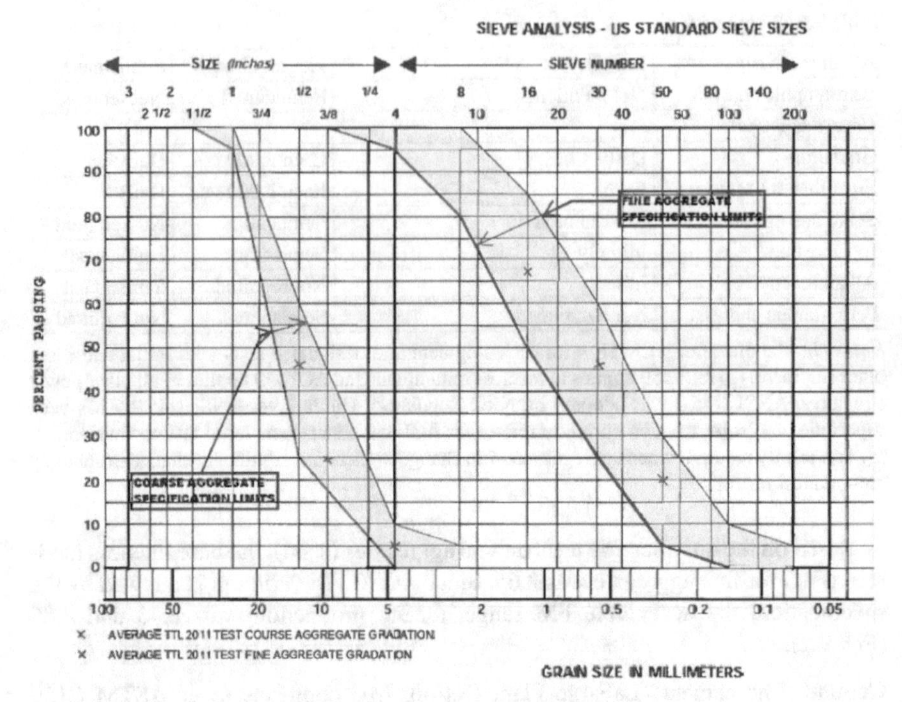

Fig. 9.2 FERMI fine and coarse aggregate ASTM C33 gradation tests

into concrete mixture to mitigate alkali reactivity and to enhance the durability of the Duct Bank concrete.

Concrete Plant Concrete was delivered from Messina ready-mix concrete plant, located in Monroe, MI, approximately 2 miles from the Fermi 2 Station. The concrete plant is operated under a quality control plan that meets Michigan DOT certification.

Transit/mixer delivery trucks used to transport ready-mix concrete were inspected in and met the Michigan DOT certification.

Table 9.7 provides selected 7-day and 28-day compressive strengths results.

Table 9.8 provides a summary cement and fly ash weight variations in individual batches.

A summary of the typical QC deviations and resolutions is listed in Table 9.9.

Aggregates Gradation Tests ASTM C136 and AASHTO T27 Comparison

ASTM C136 (Ref. [2]) and AASHTO T27 (Ref. [3]) are standard test methods for sieve analysis of fine and coarse aggregates. Table 1 of ASTM C136 and AASHTO T27 limit the quantity of material on a given sieve so that all particles have opportunity to reach sieve openings a number of times during the sieving operation, as reproduced in Table 9.10.

Table 9.7 Selected compressive strength IFSI cask storage results

Date	Section	Batch ticket	CEM + FA (LBS/YD[1])	Slump + % air	Average strength (PSI)					
					7 days	28 days	Gain	7 days	28 days	Gain
07/17/09	1	11,200,137	374	7.75	2980	4800	1820	2750	4520	1770
		11,200,151	387	9.25	2520	4240	1720			
	2	11,200,164	370	10.50	2900	4210	1310	2840	3960	1120
		11,200,187	375	11.60	2770	3700	930			
07/20/09	3	11,200,223–224	397	12.25	2180	3320	1140	2420	3630	1210
		11,200,241	377	9.90	2650	3940	1290			
	4	11,200,259	373	9.75	2720	4150	1430	2740	3960	1220
		11,200,275	375	11.70	2770	3770	1000			
07/21/09	5	11,200,290	381	9.25	2610	3580	970	2580	3600	1020
		11,200,302	372	10.60	2550	3620	1070			
07/22/09	6	11,200,334	389	9.25	2940	3840	900	2560	3680	1.020
		11,200,346	373	10.75	2300	3520	1130			
	7	11,200,368	374	10.45	2390	3600	1210	2380	3660	1280
		11,200,381	375	9.50	2370	3720	1350			
07/23/09	8	11,200,404	374	7.80	2390	3410	1020	2840	3900	1060
		11,200,415	374	10.05	3295	4380	1110			
Average			**378**	**10.02**				**2650**	**3.860**	**1210**
Mean			**375**	**9.98**				**2700**	**3860**	**1160**

Notes (1) &$$$;

Table 9.8 Cement and fly ash batch weight variances

Batch ticket		Cementitious MTL (LBS/YD²)			Water/ cement	Variance (%)		Strength 28-days (PSI)
		Cement	Fly ash	Total		Cement	Total	
July 17, section 1 concrete placement								
1	11,200,143	300	70	**370**	0.590	0	-2	
2	11,200,146	**294**	75	**369**	0.588	-2	-2	
3	11,200,148	**296**	76	372	0.587	-2	-1	
4	11,200,149	**294**	76	**370**	0.590	-2	-2	
5	11,200,150	305	86	**391**	0.559	1	+4	
6	11,200,151	**312**	76	**388**	0.563	4	+3	**4240**
7	11,200,152	**292**	82	374	0.581	−3	−1	
8	11,200,153	**289**	88	377	0.580	−4	0	
9	11,200,154	**291**	84	375	0.581	−3	0	
10	11,200,155	**292**	77	**369**	0.591	−3	−2	
11	11,200,156	301	67	**368**	0.590	0	−2	
12	11,200,157	305	64	**369**	0.589	−1	−2	
13	11,200,160	**294**	81	375	0.582	−2	0	
14	11,200,163	299	71	**370**	0.586	−1	−2	
Average strength								4520
July 17, section 2 concrete placement								
1	11,200,164–165	**309**	88	**397**	0.522	+3	+6	
2	11,200,170	**292**	78	**370**	0.590	−3	−2	
3	11,200,171	**291**	79	**370**	0.587	−3	−2	
4	11,200,172	**294**	81	375	0.582	−2	0	
5	11,200,177	**290**	80	**370**	0.590	−4	−2	
6	11,200,181	**294**	76	371	0.586	−2	−1	
7	11,200,182	299	71	**370**	0.588	−1	−2	
8	11,200,191	**294**	78	372	0.588	−2	−1	
9	11,200,194	**291**	86	377	0.581	−3	0	
10	11,200,198	**295**	81	376	0.582	−2	0	
Average strength								33,960
July 20, section 3 concrete placement								
1	11,200,223–224	301	68	**369**	??	0	−2	3320
2	11,200,237	**306**	68	374	0.581	+2	0	
Average strength								3630

Note: The ASTM C94 allowable variation is ±1% for cement and cementitious material (±1%. Of 301 lbs for cement and ± 1%. of 376 lbs for cementitious material)

For the fine aggregate, ASTM requires a 300 g minimum sample weight, and there is no upper limit, although the precision statement of the standard is based on the AASHTO test data which used 500 g test samples. USBR concrete manual (Ref. [2]) recommends the fine aggregate sample size according to FM of the material. For fine aggregate FM >2.5, the recommended test sample size is 400–1000 g.

Table 9.9 FERMI 2 ISFSI concrete pad and electrical duct bank quality control during production

Specification requirement	Deviation	Resolution
Compressive strength shall be between 3000 and 4000 psi range	5 tests out of 16 tests strength exceeded 4000 psi	Required engineer's evaluation
Batching tolerance shall conform to ASTM C94	26 batches out of 100 exceed the allowable tolerance for cement (1%)	Acceptance of the deviation was based on the 28-day compressive strength
Batch ticket shall include quantities of all concrete materials	Incomplete batch ticket	A full-time inspector at batch plant to verify compliance
Concrete shall be discharged within 90 minutes of batching	One batch is discharged 105 minutes after batching	A full-time inspector at batch plant
Minus 200 material(1.5%) in coarse aggregate	Several tests exceed the limit (range: 1.7–1.9%)	Washed aggregates for compliance. Also relaxed the limit to 2% if the fines are free of clay or shale
Fineness modulus(FM) of sand during production shall be with +/– 0.2 of the base FM	Sand gradation is such the base FM design tolerance is exceeded	Use FM as the average value based on 10 consecutive tests
Air content of concrete shall be within 1.5% of the specified value	Individual batches with inadequate or lower air than specified	Use average air content based on 10 consecutive tests for acceptance
Concrete slump shall not exceed by 1 inch of the specified value	The actual slump exceeded the specified values by up to 2 inches	Allowed concrete mixing for additional 50 revolutions to lower the slump
The total number of truck revolutions upon concrete delivery shall not exceed 300	Total revolutions are exceeded but slump and air content of the batch are within tolerances	Accepted the batch as-is
ASTM C33 limit on friable particles in fine aggregate	Friable particles in fine aggregate exceeded 7%	Allowed increase in friable particles provided the alkali-silica reactivity (ASTM C1567) expansion is <0.1%
ASTM C33 gradation limits	The amount passing the individual sieve was 2–5% higher than specified	Allowed the deviation, provided the FM (average 10 tests) was within 0.20 of the base FM

ASTM C136 recommends the use of the 8-inch and 12-inch sieves for sieving. In case of overloading of any sieve in Table 4 above, ASTM C136 requires splitting the sample into two or more portions or to insert additional sieve with opening size intermediate between the sieve that may be overloaded. ASTM recommends 12-inch sieve for coarse aggregate testing to prevent possible overloading and damage to the sieves. For aggregate samples exceeding 20 kg, ASTM recommends a mechanical sieve shaker device.

Table 9.10 ASTM C136 allowable material retained on sieves and test sample size for fine and coarse aggregates

Reference	Sieve opening	Maximum allowable material retained on sieve		Test sample size coarse aggregate	Test sample size fine aggregate
ASTM C136	Inches (mm)	8-inch diameter	12-inch diameter		
	1 (25)	1800 g	4200 g	10,000 g	
	3/4 (19)	1400 g	3200 g	5000 g	
	1/2 (12.5)	890 g	2100 g	2000 g	
	3/8 (9.5)	670 g	1600 g	1000 g	300 g (min)
	No.4 (4.75)	330 g	800 g		
AASHTO T27	1 (25)	1800 g	4200 g		
	3/4 (19)	1400 g	3200 g		
	1/2 (12.5)	890 g	2100 g		
	3/8 (9.5)	670 g	1600 g		
	No.4 (4.75)	330 g	800 g		
	−no.4 (−4.75)	200 g	400 g		
USBR concrete manual					400–1000 g (FM 2.5–3.5)

Slump and Air Content Tests Slump and air contents tests were performed at the truck discharge, every 50 cubic yards.

The slump and air content ranges were within the specified requirements. The workability of the concrete was adjusted in the field with dosage of a high-range water-reducing (HRWR) admixture, additional mixing, and testing.

Compressive Strength Tests The average 7-day and 28-day compressive strength for the 2009–2011 period were:

	7-day	28-day
For the entire duct concrete	5005 psi	6731 psi

For the entire concrete placement, the minimum 28-day compressive strength was reported to be 5200 psi. The average 28-day compressive strength for the non-conforming tests was 40% higher than the required design strength (4000 psi). The deviations in testing, reported in Table 9.9 were, therefore, accepted based on the compressive strength results.

Table 9.11 Concrete Mix 4000F-3 field tests summary

	W/C ratio	Slump	Air content	Unit weight	Concrete temperature	Compressive strength psi	
		Inches	%	Lb/cf	Degrees F	7 days	28 days
Minimum	0.42	4.00	2.8	145.2	47	3120	4250
Maximum	0.47	9.00	7.2	151.0	63	3975	5445
Average		7.00	4.6	148.2	58	3472	4967
Tests		19	19	19	19	19	19
Standard deviation,							410

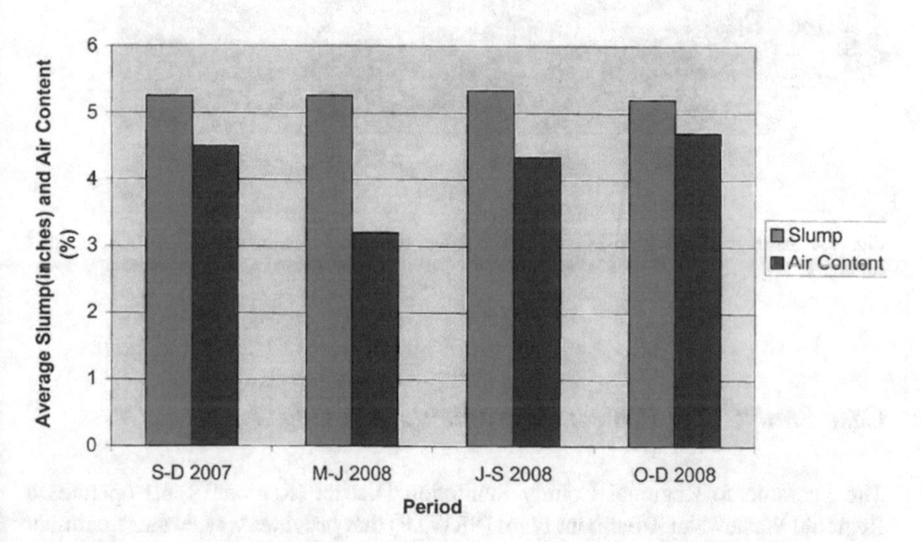

Fig. 9.3 Production slump and air content for Concrete Mix 4000F-3 tests summary

Results Based on a review of the concrete placement data at Fermi 2 Station, the 28-day compressive strength of concrete for Fermi 2 Electrical Duct Bank met and exceeded the acceptance criteria of the specification that "concrete shall be considered satisfactory if compressive strength results on all three (3) 28-day specimens or both 56-day specimens equaled or exceeded 4000 psi."

Case Study: National Enrichment Facility, Lea County, NM

This testing of concrete mix designs for National Enrichment Facility, Lea County, New Mexico, was performed during January–April 2007 period. Mix proportions for Class Mix 4000F-3 (4000 psi) were selected and tested for slump, air content, unit weight, and strength, as shown in Table 9.11 and Figs. 9.3 and 9.4.

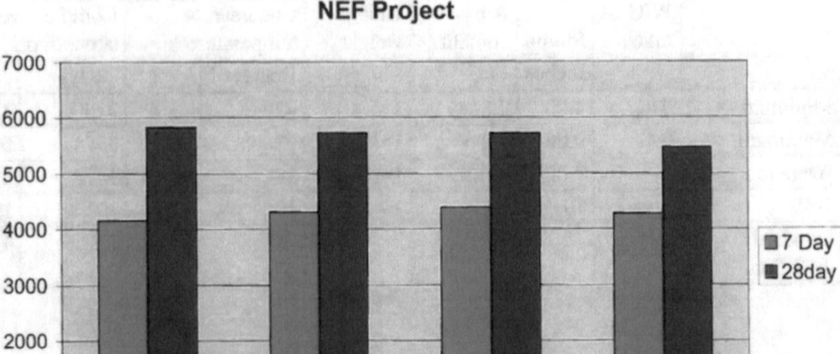

Fig. 9.4 Average 7-day and 28-day compressive strength for Concrete Mix 4000F-3 tests summary

Case Study: TTF Project, Sacramento, CA (Fig. 7.1)

The Sacramento Regional County Sanitation District (Regional San) operates a Regional Wastewater Treatment Plant (SRWTP) that provides wastewater treatment to the Sacramento area and surrounding cities, serving approximately1.3 million customers. SRWTP currently uses a secondary treatment process capacity of 181 million gallons per day (mgd) average dry weather flow (ADWF). The treated effluent is discharged into the Sacramento River.

Regional San has embarked on a comprehensive upgrade. The upgrade, called the ECHO Project, includes tertiary processes, filtration, and enhanced disinfection, consistent with recycled water requirements under Title 22, Division 4, Chap. 3, of the California Code of Regulations or equivalent. The filtration would provide up to 217 million gallons per day of effluent.

Begun in 2013, the program has a 12-year schedule and an estimated construction cost of $1.735 billion. Program is focused on assuring plant capabilities well into the future and improving waste in the Sacramento River, CA (under construction).

TTF Concrete Mixtures for Durability A HPC, with a shrinkage reducing admixture (SRA) and a permeability-reducing admixture (PRA) and 4500 psi compressive strength at 28 days, is being used for the project. The performance requirements for the TTF liquid-retaining HPC and concrete mixture proportions are

Table 9.12 TTF mixture 7-day and 28-day compressive strength (November 2018–June 2019) data

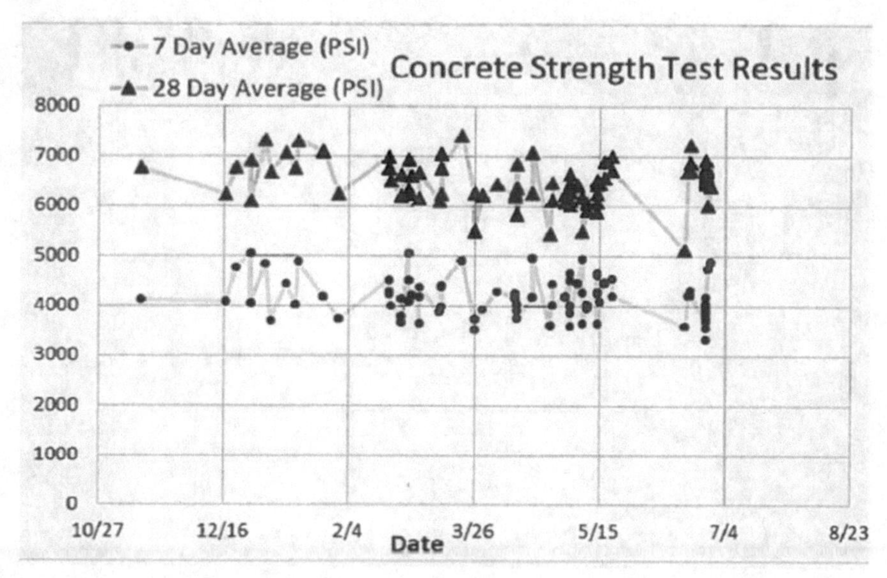

discussed in Chap. 7. The 7-day and 28-day compressive strengths results from November 2018 through June 2019 Period are shown in Table 9.12.

(average 28-day strength = 6473 psi; standard deviation = 426 psi)

References

1. ASME NQA.1 Quality Assurance Requirements for Nuclear Facility Applications
2. ACI 359 Code for Concrete Reactor Vessels ans Containment.
3. Concrete Manual, United States Department of Interior, Bureau of Reclamation, Eighth Edition.
4. Cavitation on Chutes and Spillway", USBR Engineering Monograph No. 42, April 190

References

Chapter 10
Concrete Repairs

10.1 General

Methods used for repair and maintenance of concrete are documented in Concrete Manual [1], published by USBR.

Proprietary Grout Product: Manufacturer's product data for manufactured proprietary grouts and mortars. Manufacturers should have 5-year success record.

Repair Drawings: Drawings or sketches that indicate the anticipated location and extent of each area to be repaired. After completion of repairs, submit as-built drawings showing the actual location and extent of repairs.

Sand: Sand for repairing concrete shall pass a No. 16 screen and shall meet the following gradations:

An epoxy bonding compound (e.g., Sikastix 360, Sikadur 370, Sikadur Hi-Mod, manufactured by Sika Chemicals Inc., or approved equal products) should be used as a bonding medium, when the depth of the area is 3 inches or less and the repair will be of appreciable continuous area. Note the pot life of the epoxy bonding compound is short and dictated by the ambient temperatures. It must be applied per the manufacturer's instructions for effectiveness.

Anchors, keys, or dovetail slots shall be provided wherever necessary to attach the new material securely in place.

Exposed reinforcing steel during chipping: Rebar "exposed" in the area of the void or honeycomb during the removal of unsound concrete should have the concrete removed to provide 1-inch clearance around the bar.

As soon as practical after removal of forms, unsatisfactory concrete shall be cut out and replaced with new concrete, with the following modifications. Surface of prepared voids shall be wetted for 24 hours immediately before the patching material is placed, unless epoxy bonding agent is used.

N. Hasan, *Durability and Sustainability of Concrete*,
https://doi.org/10.1007/978-3-030-51573-7_10

10.2 Repair Methods

Cosmetic Repairs Small size holes having surface dimensions about equal to the depth of the hole (for holes left after removal of form ties, grout insert holes and slots cut for repair of cracks) should be accomplished by dry-pack filling. Hole shall be clean and sound with no loose or powdery material. Use high-modulus epoxy.

Structural repairs (areas equal to or greater than 1 square foot in surface area and 4 inches in depth or any area with rebar exposed) should be accomplished by one of the following methods:

Replacement Concrete The same of higher strength concrete may be used to make repairs. Concrete replacement should be used when the defect extends entirely through the concrete section.

Replacement Mortar In repair and replacement of unsatisfactory concrete in areas where repair with replacement concrete or replaced aggregate is not possible to assure a satisfactory patch (i.e., ceilings, underside of beams, upper sides of walls adjacent to ceilings, etc.), the following repair methods will be used.

Poured Replacement Mortar: Replacement mortar should be used in areas where location, congestion, or size prohibits the use of dry pack. However, unlike pressure grouting, replacement mortar may be poured into place provided that the hole configuration allows for natural venting of air. Hole configurations which will entrap air may not be repaired using poured replacement mortar. All unsatisfactory concrete shall be removed, and an epoxy bonding compound shall be applied before repairing. The edges of all voids shall be dovetailed to aid in attaching the grout securely in place:

1. Chip out the concrete to remove all the voids, honeycombs, loose dust, and other loose materials.
2. Brush onto the existing concrete surface a grout bond coat. This can be a mixture of san passing the #30 sieve, cement, and water in accordance with Reference 2.1. Then place a tongue and groove board, 12 feet ± wide by length of the repair area over the beginning of the void, such that the board overlaps onto the existing smooth concrete at both ends and on one side. The board shall be securely placed and shall be in firm contact with the smooth concrete, surrounding the area to be repaired. Pack grout thoroughly in the space between the board and the shipped out concrete surface. The repair material shall be placed parallel to the tongue and grooved board and in a horizontal direction. Check the board and ensure against bulging.
3. For large areas of repair, it may be necessary to make the repair by applying the repair material in layers. When the size of the area to be repaired requires layering, it is important to allow the previous layer to reach initial set before brushing on the grout bond coat and continuing with the next layer.

4. Proceed as above until less than the width of one tongue and groove board remains in the area to be repaired. Prior to placing the last board, pack the repair material thoroughly as in step #3, and then quickly place and secure the board.
5. Cement and concrete quality sand shall be separately bagged up and weighed to assure accurate quantities. Water will be added by concrete workers in the field to meet the consistency requirements of dry pack.
6. Preplaced aggregate may be used for holes extending entirely through the concrete section, for areas greater than 1 square foot and deeper than 4 inches, and holes in reinforced concrete which are greater than one-half square foot and which extend beyond the reinforcement. The repair shall be made by making a complete filling of the void with washed gravel or broken stone and liquid. Portland cement grout shall be placed through filler pipes under pressure. Pipe nipples shall be placed through the forms at the bottom of the void so that the grout rises upward through the aggregate to spill through a vent at the top edge of the void.

Method of concrete repair around waterstops using polyurethane adhesive (Sikastix 360, manufactured by Sika or approved equal).

For repair of concrete due to poor consolidation at the waterstop, the following steps are recommended. Refer to figure below.

1. Saw cut and remove defective concrete at the joint.
2. Cut and remove the waterstop, leaving two 12-inch pieces at the ends.
3. Thermally fuse the new waterstop.
4. Form and place epoxy non-shrink grout, using an epoxy adhesive for bonding.
5. Moist cure for 3 days.

Saw Cut and Splice

When damage to an embedded waterstop, residing in a moving joint, is extensive and extends near to or beyond the concrete face, removal of the concrete and damaged waterstop is often the only remedy.

Remove concrete, full depth, surrounding the damaged area and extending 12" longitudinally past the damage and 2" past the waterstop edge. Cut and remove the damaged section of waterstop, leaving two 12" free ends of waterstop. The free ends will be necessary to facilitate welding of the new splice section of waterstop. Thermally fuse the splice section of waterstop in position with a thermostatically controlled, Teflon coated heating iron. Form and place a non-shrink epoxy grout in the area of the removed concrete. A neat epoxy coat will help insure a good bond to the existing concrete. Take measures to insure good consolidation of the grout around the waterstop.

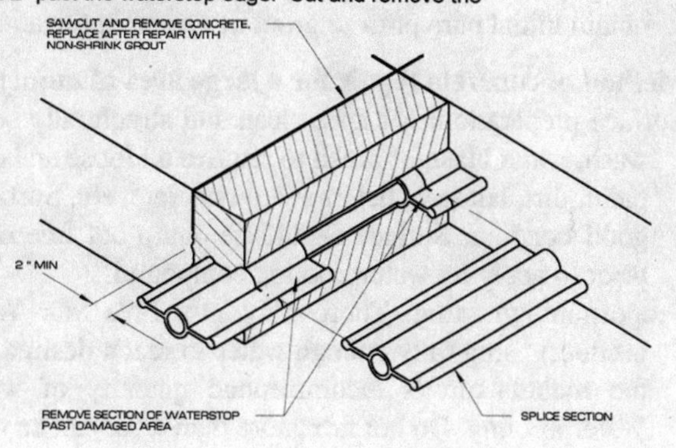

Crack Repairs Chip out crack in a "V" shape with a minimum of 1/4 " depth and a 1–1/2" surface width. Concrete must be clean and sound. It may be dry, damp, or wet but free of standing water. Remove dust, grease curing compound, waxes, foreign particles, and disintegrated material by mechanical abrasion methods.

Proportioning/Mixing: Mix components A and B as per manufacturer's recommended proportions into a clean pan. Mix thoroughly for 3 minutes with paddle on low speed (400 to 600 rpm) and slowly add up to 1 part by loose volume of Colma Quartzite Aggregate or Kiln Dried Sand (sandblasting sand) and continue to mix to uniform consistency.

To mix Sikaset 517, put a small quantity of fresh, dry Portland cement in rubber cup or flexible blow and add Sikaset. Mix quickly with round-ended trowel. Sikaset 517 plugs set in 45–60 seconds at 70 degrees F.

Application: For slight seepage apply Sikastix 360 with a spatula, trowel, or caulking gun until level with existing concrete surface. Do not apply in layers greater than 1–1/2 inches thick.

For moderate to heavy seepage, always seal slower leakage first. Then cut a deep (2–3 inches) conical hole into leaking concrete following course of water. When mixed Sikaset 517 thickens, form into carrot-shaped plug, and push firmly into prepared hole. Hold fast until set. Set will be slower in cool weather and faster in warm weather.

Curing: The patched area shall be cured by covering the said area immediately by an approved nonstaining water-saturated material which shall be kept wet and protected against the sun and wind for a period of 12 hours. This area shall be kept wet continuously by a fine spray or sprinkling for not less than 7 or 14 days. As an alternate to set curing, after the initial 12-hour wet cure, an owner-approved curing compound may be used. The curing compound shall be compatible with and by the same manufacturer as the surface and topcoat unless the curing compound is to be removed prior to coatings application.

When multiple coats are required, initial cure for each layer should be reached before additional coats are applied.

For large areas of repair at waterstop, Sikastix 360 shall be applied and allowed to obtain initial cure prior to grout or mortar application.

Method of concrete repair for a large area of numerous water seepage cracks:

Surface preparation: Must be clean and structurally sound. Chips, sandblast, acid wash, water blast, or brush to remove all loose and deleterious materials such as paint, dirt, laitance, deteriorated concrete, etc. Surfaces must be rough to insure good bonding. Surface should be damp but free of excess water immediately prior to applying waterproofing compound.

Proportioning/mixing: Thoroughly mix Five Star Waterproofing (or equivalent product) using only enough water to reach desired consistency. Do not exceed the manufacturer's recommended quantity of water per bag of Five Star Waterproofing. Do not mix more than what can be placed in 30 minutes.

Application: Five Star Waterproofing filling all voids, holes, and pores. If desired, Five Star Waterproofing can be easily dressed using a trowel, sponge, or wood float.

Curing: The surface should be cured by keeping the affected area continually moist during the first hour after application. All surfaces should be kept above 50 degrees F during application and curing.

10.3 Quality Assurance

1. Determine the appropriate method of structural repair to be used after form removal if structural repair is necessary.
2. Initiate a concrete repair report for the structural patch(es).
3. Complete the pre-checkout section of the placement report.
4. Monitor work operations to ensure compliance with design documents, specifications, standards, and codes, and sign off the design approval slot on the repair report.
5. Inspect and document repair areas in accordance with applicable construction inspection (CI) procedures.
6. Monitor cure and protection of repair areas in accordance with applicable procedures.
7. Ensure that the necessary safety precautions are taken when using the Sikastix 360.

Case Study: Foote Dam Rehabilitation at Au Sable River, MI (Ref. [2])

The Foote Hydro Station, located on Au Sauble River, 9 miles upstream of Lake Huron in the upper Michigan Peninsula, is a 9 MW hydroelectric facility that was constructed in 1918. The 75-year concrete, which was not air-entrained, was found to be deteriorated at or above the water line of the dam, with cracks in portions of the piers, caused by the freezing and thawing damage.

In 1993, repairs were performed to restore the deteriorated water passages including the spillway, piers, log-chute, and abutment wall. The reconstruction was accomplished in less than 12 months and included the following: removal and salvage of existing gate hoists and gate controls; removal of access bridge; removal of tainter gates; installation and underpinning of log chute wall; surface demolition of abutment wall using hydro-demolition with 20,000 psi pressure. Reference [2] describes the details of the Foote Dam Spillway rehabilitation, with highlights presented herein.

NDT (ASTM C805) Schmidt hammer: During demolition, the depth of unsound concrete was identified by use of ASTM C805 method, with readings of less than 3000 psi. The unsound concrete in the 4-ft wide piers was excessive, resulting in complete removal, downstream of the gate.

Concrete for repairs included a low-alkali cement since the aggregate was found to be reactive. An air-entraining admixture was added to the replacement concrete.

Non-destructive Testing (NDT) An NDT program, consisting of transient dynamic response (TDR) method was conducted to verify the soundness and uniformity of concrete in the inclined apron slab, based on dynamic stiffness. The spacing of TDR tests was performed on 2.5 ft. by 5 ft. grid spacing. The TDR values for the apron concrete ranged from 1.32 to 2.64 kips/in, which were considered acceptable.

The ultrasonic pulse velocity (UPV) method, in accordance with ASTM C597, was applied for the main and intermediate piers below the inclined apron. The UPV values for the main piers were acceptable (range 13,000–5000 ft/sec), while for the intermediate piers were variable (range: 6000–12,000 ft/sec) and indicated variation in the surface conditions and irregularities and not due to integrity of concrete (Figs. 10.1, 10.2, 10.3, and 10.4).

Fig. 10.1 Demolition of piers and left abutment after the erection of a cofferdam (consisting of sheet pile (PZ 35) panels and horizontal wale and steel supports at the main and intermediate piers)

Fig. 10.2 Reconstruction of log chute wall

Fig. 10.3 Formwork for log chute walls

Fig. 10.4 Reconstruction of piers and left abutment wall. Partial removal of right abutment wall (after hydro demolition) is visible

Conclusion The successful repairs of the spillway illustrated that the structural rehabilitation is a proven and viable alternative in extending service life of a small hydro plant.

Case Study: Repairs to Hebgen Dam, MT

Installation of Controlled Low Strength Materials (CLSM) for Underwater Applications

Background A controlled low strength material (CLSM) as temporary fill for the rehabilitation of an existing intake structure (where dewatering was not feasible), was installed at the Hebgen Dam Intake Rehabilitation, Montana. The enhanced CLSM included admixtures to facilitate tremie placement, under low ambient air and material temperatures and water depths exceeding 60 feet. The enhanced CLSM was used as a Upon rehabilitation of the intake structure, the 200-psi strength CLSM was excavated using conventional equipment.

The CLSM mixture was designed to produce a self-leveling mixture for placement with minimal wash-out, high 24-hour strength to allow subsequent lifts every day, and a low 28-day design strength to allow future excavation and removal. The CLSM trial mixture included a combination of four different admixtures including anti-washout admixture, high-range water-reducing admixture, air entraining, and accelerating admixture to meet the unique design requirements. The admixture dosage rates and dispensing sequence were verified during the trial batches.

General The flowable controlled low strength material (CLSM) for filling the interstitial space between the existing intake structure and temporary cofferdam was designed to meet the following objectives:

1. Displace water and produce buoyant initial weight near that of water (this eliminated the staged dewatering requirement for the intake rehabilitation).
2. Maintain its self-leveling characteristic with minimal washout.
3. Produce early 24-hour strength to allow subsequent lift.
4. Meet the design low strength (250 psi at 7 days) to allow future excavation and removal.

Controlled low-strength material (CLSM) is a self-compacted, cementitious material primarily used as a backfill in place of compacted fill. CLSM (also known as flowable fill) is a cementitious material manufactured using cement, fines (fly ash, sand, or native granular materials), and water mixed in a concrete plant or a truck mixer. The combined ingredients produce a flowable mixture that can be installed without compaction. The CLSM, due to its flowability and low placement costs, is being increasingly used for structural backfill, pavement bases, subbases and subgrades, bedding for pipes, electrical and conduits, erosion control and nuclear facilities, and waste disposal sites, where construction schedules are critical and the conventional compacted backfill methods pose delays.

The CLSM mixtures with compressive strengths up to 200 psi are considered excavatable using hand equipment, but at higher strength, the CLSM requires heavy equipment (a backhoe) for excavation. American Concrete Institute Report ACI 229 R [3] provides guidelines for CLSM applications, including materials, mixing, placing, and quality control requirements.

Reference [4] describes the use of underwater CLSM for encapsulating and filling the high-level waste tanks at Savannah River Complex, Aiken, SC. Reference [5] describes the use of CLSM as an alternative to soil/structural fill and base course beneath the floor slabs of National Enrichment Facility (NEF) in New Mexico. The CLSM mixture included Portland cement, fine aggregate, with 30% material passing No.200 sieve, and air entraining admixture. The mixture was mixed in a truck mixer and designed to produce a flowable mixture for placement without compaction. The compressive strength was 100-200 psi at 28 days, using a site developed testing procedure.

CLSM Mix Proportions CLSM mixture proportions were established from trial batches performed in the laboratory, and on-site, using cement, fly ash, aggregates, water, anti-washout admixture (AWA), high-range water-reducing admixture (HRWRA), air-entraining admixture (AEA), and accelerating admixture (AA). The following admixtures, manufactured by BASF Corporation, were considered:

AWA: Rheomac 450 (BASF) to minimize washout
AEA: Rheocell Rheofill (BASF) to maintain 20% air content
HRWRA: Glenium 3030 (BASF) to maintain flowability
AA: Pozzolith NC 534(BASF) to accelerate setting time

Table 10.1 CLSM mixture proportions (pounds per cubic yard unless noted)

Component	Unit	Mix A note 1	Mix B note 1	Mix C note 2
Cement (Type I/II)	lb/cy	150	150	300
Fine aggregate	lb/cy	1155	1540	2650
Coarse aggregate (3/8″)	lb/cy		473	None
Fly Ash	lb/cy	500	500	None
Water	lb/cy	425	600	240
Water/cementitious ratio		0.70	1.0	1.0
Air-entraining admixture/Rheocell	fl. oz./100 lb	3.5	1.5	
Anti-washout admixture	fl. oz./100 lb	15	30	15
HRWRA, Type F	fl oz./100 lb	3.0	–	3.0
Accelerating admixture, Type C	fl. oz./100 lb	40	–	40
Slump flow (ASTM C1611)	Inches	24	28	18
Air content (ASTM C231)	Percent	9.8	4.7	18.0
Washout (CRD-61)	Percent	4.4	6.7	9.5
Unit weight (ASTM D6023)	lb/cft	123.5	120.5	108.50
Compressive strength (ASTM D4832)				
24 hours	psi	40	20	105
7 day	psi	220	95	240

Note 1: Based on laboratory tests results (3 tests)
Note 2: Based on field production tests (5 or more tests)

The preliminary trial mixes were performed in a laboratory to assess the flowability and washout behavior of two CLSM Mixes, Mix A and Mix B, both including minus 3/8 inch coarse aggregate. Mix B, containing a higher w/c ratio (1.0) and run without the accelerating admixture (NC-534), exhibited a longer setting time and lower 7-day strengths than Mix A. A trial Mix C, replacing the fly ash and 3/8 inch aggregate with the increased cement content and fine aggregate, at a w/c ratio of 0.8 exhibted higher strength.

Table 10.1 presents the mix proportions, admixtures and dosages, and test results for the CLSM Mixes A, B, and C.

Mix CLSM-C was evaluated for excavatability using a test section. The test section was excavated after 7 days age, using a 12-inch and 24-inch buckets. The results were satisfactory; however, 12-inch bucket was found to be more effective, as shown in Fig. 10.5.

CLSM Installation Tremie placement of CLSM was made by the contractor in accordance with a work plan, which included the following features:

1. A 6-inch tremie pipe, hung from a crane, was placed at the center of each lift. The tremie pipe was marked in 3-foot increments for controlling the lift height during placement.
2. Divers were employed to monitor the flowability and lift height control and to observe/mitigate washout/leakage during placement.

Fig. 10.5 CLSM excavatability test

3. The tremie pipe was relocated to a new spot, approximately 10 feet from the previous location, after the 3-ft mark was reached.
4. The time interval between lift was 12 hours minimum.
5. No surface preparation was required between successive lifts.

The vertical face of the intake structure stoplogs was sealed using 40-mil polyethylene sheeting, sand filled bags, and concrete blankets, to minimize leakage and loss of CLSM during installation.

CLSM was batched on site using a mobile batch plant and delivered by two concrete trucks. CLSM materials were batched at a central batching facility to assure uniform quality and consistency of mixture between successive batches. The portable central batch plant, meeting the NRMCA certification requirements in accordance with ASTM C94, was used. The various admixtures were dispensed in accordance with the manufacturer's recommendations. First cement, sand, water, and Rheocell (air-entraining admixture) were added to the batch. Then, accelerating (NC 534) and HRWRA (Glenium 7500) admixtures were then added in sequence, each mixed for 5 minutes. Finally, AWA (UW-450) was added, and the batch was mixed for additional 5 minutes, before discharging into the pump hopper.

CLSM was installed between the cofferdam and the intake structure in 3-foot lifts using a pump truck (Fig. 10.2) and 6 inches tremie pipe (Fig. 10.3) that extended, from 26 feet to 67 feet in water, to the bottom of each placement. A pig was used in the pipe to prevent segregation. The lift thickness was confirmed by sounding with a weighted measuring tape.

CLSM Quality Control CLSM strength specimens should be placed into 6 x 12-inch, pre-split and taped cylinder molds without rodding or tapping, in accordance with ASTM D 4832. The cylinders should be slightly heaped off above the molds to allow for the initial settlement and left undisturbed for 24–36 hours after casting at the site testing agency laboratory. The cylinders should be tested

Fig. 10.6 Delivery of CLSM and pump Setup

Table 10.2 CLSM production test summary

Date	W/C ratio	Slump flow, inches	Air content %	Conc. temp F	Air temp F	Unit weight lb./cft	Wash-out %	7-day strength psi
11/2/10	0.85	12.0	22	46	42	105.5		310
11/4/10	1.03	18.5	18.5	59	38		9.3	
11/5/10	1.03	15.75	18.5	59	50			150
11/8/10	0.99	19.25	19.0	44	33	108		
11/10/10	0.99	18.87	18		33	106		
11/11/10	0.99	18	16	49	30	112		
11/13/10	1.07	19.25	Note 1	51	33	107	6.6	170
11/15/10	1.03	19.5	Note 1	51	36	111		

Note 1: Air meter not holding air

Fig. 10.7 Discharging CLSM into pump hopper

Fig. 10.8 Dewatering after CLSM installation

for compressive strength at 7-day and 28-day or 56-day ages. At the required age, the specimens should be carefully removed from the molds, allowed to dry at room temperature for 4 hours, and the ends of specimens are ground prior to testing. Since CLSM specimens may be fragile at early ages, 1-day and 3-day strength tests are not recommended.

A total of 983 cubic yards of CLSM was successfully placed between November 2 and November 17, 2010 (Fig. 10.6). Table 10.2 presents quality control tests performed during production including slump-flow (ASTM C1611), air content (ASTM C231), unit weight (ASTM D6023), and compressive strength (ASTM D4832) tests (Figs. 10.7, 10.8, 10.9, and 10.10).

Fig. 10.9 Final lift of CLSM exposed upon dewatering

Fig. 10.10 Existing cofferdam and intake structure after dewatering

References

1. Concrete Manual, A manual for the Concrete Construction, United States Department of Interior, Bureau of Reclamation, Eighth Edition
2. High Performance Zero-Bleed CLSM/Grout Mixes for High Level Waste Tank Closures Strategic Research and Development- FY 99 Report (WRS-RP-01014, Rev 0), Chistine A Langton, Westinghouse Savannah River National Laboratory, Aiken, SC 29808, and N Rajendran, Consultant, Marrietta, GA 30067, January 15, 2000
3. Applications of Controlled Low-Strength Materials (CLSM) for National Enrichment Facility, Eunice, New Mexico, by Nausherwan Hasan, URS Corporation, New York, NY 10279 and Terry Gardner, Consultant, Texas, presented at the ASCE 2009 Pipeline Conference, San Diego, CA
4. Hasan, Nausherwan, and Sowers, Donald, "Foote Spillway Rehabilitation", International Water Power Conference, V, Procedings, Nice, 1995
5. ACI 229R-99, Controlled Low-Strength Materials, Reported by ACI 229 Committee

Chapter 11
Durability Requirements for 100-Year Service Life

11.1 A Historical Perspective of the Twentieth Century

Looking forward to the concrete durability and sustainability challenges for the 100-year service life, an overview and historical perspective of concrete structures built in the USA during the last century may be helpful. In Ref. [1], Timothy Dolen provides a historical development of concrete for the US Bureau of Reclamation (USBR) infrastructure across the 17 western states from Arizona to Montana, over the last 100 years. USBR has been a premier institution in addressing the numerous challenges, and providing solutions, to constructing durable concrete in the twentieth century. These issues included the basic concrete materials, methods of batching, mixing, placing, and protecting concrete, as well as environmental deterioration mechanisms, including freezing-thawing cycles and acidic and chloride environment. Reference [1] lists three of the most critical natural deterioration mechanism affecting the reclamation structures:

Sulfate Attack: First observed on Sun River project in Montana in 1908.

Alkali-Aggregate Reaction: First experienced on Falls Dam in Ohio in 1940. This led to petrographic techniques to identify reactive aggregates, long-term testing, and the requirement of a 0.6 percent alkali limit for cement in 1941.

Freezing and Thawing Deterioration: Spillway training wall at Lahontan Dam (constructed in 1915). This led to the development of air-entrained concrete by 1942.

By the end of World War II, USBR had overcome the above primary causes of concrete durability problems in the west.

The other contributions to concrete durability during the post-war generation were new additives to achieve greater economy and durability:

1950: Large-scale use of fly ash to improve sulfate resistance of concrete, improved concrete workability, and decreased porosity of the cement paste
1980: US Conservation Recovery Act, reuse of recycled materials including fly ash in concrete
1980s: Use of superplasticizers to improve workability of concrete, leading to self-consolidating concrete (SCC)
1980's: Use of silica fume to reduce porosity and permeability of concrete

N. Hasan, *Durability and Sustainability of Concrete*,
https://doi.org/10.1007/978-3-030-51573-7_11

The innovative developments and rapid implementation of the above advances kept the concrete industry at the forefront of the state of the art through the twentieth century. The concluding remarks of Mr. Dolen [1]: "Reclamation must now face the critical task of maintaining the existing infrastructure to meet the needs of the twenty-first Century. The aging of concrete structures will require a major investment for continued operation. The most immediate needs are to protect concrete constructed before the big three durability issues were solved. Unfortunately, this only narrows the field down to about the 50 percent of our inventory constructed before the World War II.A decision support information on their long-term, service life potential. With this information, Reclamation intends to present the status of our concrete infrastructure on a time-line to prioritize funding for protection before deterioration processes damages these facilities beyond repair."

11.2 Achieving a 100-Year Service Life

Historically, as noted in Ref. [1], it was not uncommon for structures to remain operationally serviceable or economically justifiable for more than 50 years in many cases. Today's new standard of incorporating long-duration operational service life into the design of monumental civil structures has brought about new challenges, which should be managed using a holistic approach to durability. Figure 11.1 shows the overall considerations that should be taken into account when determining the service life. A holistic approach to durability involves the merging and assimilation of design, construction, and operations in accordance with the provisions found in the technical requirements for the project. The project team requires the services of a qualified concrete laboratory, with extensive experience in durability engineering to identify solutions for achieving a 100-year service life for a project.

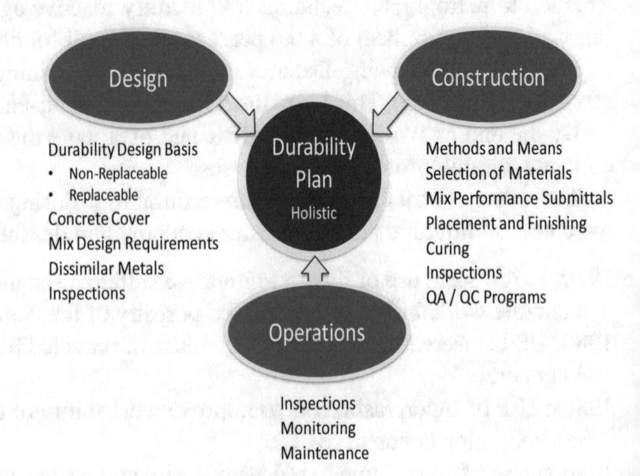

Fig. 11.1 Holistic approach to the durability plan

11.3 Quality of Materials and Design

A holistic durability plan should include the design phase, which should take into consideration reinforcement selection and location, mix design requirements, and design detailing. The design phase should be completed on a per element basis to ensure that each element design is optimized for its specific exposure. The service life of each element should be verified through the use of probabilistic modeling which allow for natural variances in the modeling parameters (Fig. 11.2).

Materials Selection It is very important to holistically consider the selection of materials for a concrete mix. Not only must the selected materials combine to achieve strength and chloride ion resistance standards, they must also not cause any harm to the concrete and protect it as required by the exposure conditions. Considerations should be made during the materials selection process to avoid degradation mechanism of concrete and identifying aggregates that are potentially susceptible to deleterious ASR, heat of hydration for mass concrete, and utilization of proper cementitious materials for sulfate resistance.

Mix Prequalification Once the materials have been selected, the mix should undergo a prequalification process. During the prequalification process, the proposed mix is batched and tested to verify it meets the durability design requirements. The test results are then included in a durability design report that should as a minimum address the following: the mix design being tested, compressive strength, shrinkage, thermal coefficient, freeze-thaw resistance, abrasion resistance, sulfate resistant, concrete cover, concrete transport properties (including ion diffusion coefficient), exposures considered, production and curing techniques, and the predicted service life.

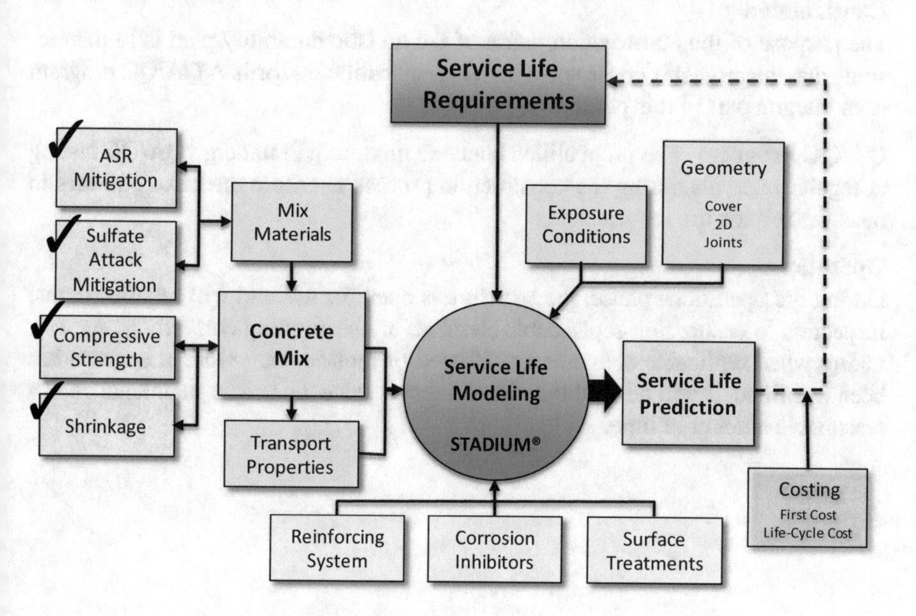

Fig. 11.2 Service life modeling flow chart

Cast-in-Place Concrete For the concrete to achieve a 100-year service life, the water-to-cementitious materials ratio (w/cm), constituent material type, volume stability (autogenous shrinkage), chloride ion diffusivity, and other requisite parameters should be considered during the selection of the appropriate mix requirements for each specific element. The requirements set for each element should be in compliance with the technical requirements and have a w/cm that should take into account the adverse effects of autogenous shrinkage on early age cracking. Also the ion diffusivity of the concrete at 28 days (or 56 if fly ash and slag are used) should take into account soil and groundwater composition. The design team should detail minimum requirements for each element in order for the contractor to pick a specific mix they are familiar with during the construction phase of the project.

Prestressed Concrete Prestressed concrete should follow the same process as cast-in-place concrete for selecting an appropriate concrete mix. However, the structures with permanent prestressing beams/girders should also be designed to receive no tensile stress in the concrete in any serviceability limit state combination to ensure the structure remains crack free.

Reinforcement The appropriate steel reinforcement should be considered for each element based on concrete cover, concrete properties, and exposure conditions, to allow for the use of ordinary carbon steel, low chrome steel conforming to ASTM A1035, or stainless steel Grade S24000/S24100 conforming to AASHTO MP 18/18 M. Furthermore, the use of calcium nitrite inhibitor as a means of raising the corrosion threshold and extending the time to corrosion should be evaluated.

Construction

The purpose of the construction phase of the holistic durability plan is to demonstrate that the project is constructed to the specification set forth. A QA/QC program is an integral part of this phase.

QA/QC Program The prequalified concrete mixture will undergo QA/QC testing at regular intervals during the construction process to ensure the mix continues to meet the service life requirements.

Operations

During the operations phase, the structure is open for use and will require routine inspection to ensure non-replaceable elements are performing correctly along with noting when replicable elements are in need of replacement. Once a concern has been identified, it will be vital that the element receive the proper maintenance in a reasonable amount of time.

11.4 Durability Modeling for 100-Year Service Life

Durability modeling combines materials engineering expertise and computer simulations to optimize the reinforced concrete system to achieve a service life goal. Typically, this is an iterative process. Each concrete element is evaluated separately based on the exposure conditions with the goals of achieving the required service life.

Probabilistic modeling (finite-element modeling) is recommended to validate the service life of the different elements of the proposed structure. Service life calculations should be performed using a finite-element modeling (such as STADIUM® or similar probabilistic modeling approaches) dedicated to the durability prediction of concrete exposed to environmental contaminants.

The model should consider the chemical composition of local cements and supplementing admixtures such as silica fume, fly ash, and ground-granulated blast-furnace slag. The model should also consider the complex interactions between the contaminants penetrating the porous network of concrete and the hydrated phases of the cement paste, including the impact of temperatures, transport of chemicals and moisture, wetting and drying cycles, and chloride penetration rate within the reinforced concrete.

Probabilistic modeling should minimally take into account the intrinsic variability of the following parameters: chloride threshold for corrosion initiation, materials properties, and concrete cover. Variability parameters of concrete properties should be based on values recorded for field mixtures found in laboratory's database, while the parameters of the chloride threshold value should be established on the basis of results of previous laboratory studies and available data, thus assuring the latter complies with project requirements.

All testing should be performed by a qualified laboratory and in particular testing for finite-element modeling for service evaluation.

11.5 Typical Scope of Work for 100-Year Service Life

Phase 1 Review existing data and research to evaluate the proposed concrete mixes for various elements for service life prediction, based on probabilistic modeling and anticipated exposure conditions.

Develop recommendations for materials and admixtures, proportions, concrete cover, and mechanical and durability properties to achieve the 100-year service life; and select appropriate geometry and steel reinforcement for cast-in-place concrete, precast, and prestressed concrete segments.

Phase 2 Perform validation testing on proposed concrete trial mixes prepared in the laboratory and at the project site; prepare the durability reports.

Phase 3 Support QA/QC testing during construction.

Concrete Durability Assessment According to the modeling approach selected, the concrete properties required to achieve the desired service life must be established right from the start, based on environmental conditions, steel type, and element geometry.

Service Life Modeling: Incorporate the exposure conditions, concrete durability assessment, and alternative reinforcing steel options in modeling simulations to develop solutions for the 100-year service life for the structures. The concrete properties without cracks should be used. The model should be representative of the real structure, including exposure conditioned, such as deicing salts and/or brackish saltwater. If applicable, different exposure zones must be considered for a same element (atmospheric, splash, tidal, buried, etc.) as part of the model. Each element should be designed for the harshest exposure condition.

Phase 1: Review of Documentation and Theoretical Calculations
In this first phase, a review of the available documentation to determine the characteristics of the concrete is made. These characteristics include the geometry, concrete cover and concrete specifications, available drawings, and the preliminary concrete schedule. The information on the applicable exposure conditions and the related degradation mechanisms should be included.

A typical concrete durability matrix is shown in Table 11.1. The first two mechanisms are related to the proper mixture design and construction. The third can be modeled to determine the time-evolving degradation with respect to applicable end of life and to evaluate if each element achieves the required service life.

The qualified laboratory should assist the design engineer in the selection of materials that will meet the project specifications pertaining to durability and general materials requirements while allowing the production of economical and durable concrete mixtures. The laboratory should evaluate the required concrete characteristics based on theoretical considerations. If deemed relevant, modifications to the preliminary concrete schedule should be considered, including alternative mixtures design.

Table 11.1 Concrete durability matrix

Type of degradation	Example	Approach
Deleterious internal reactions	Alkali-aggregate reactions,	Select aggregates and cements to mitigate degradation by ASR in concrete
Physical mechanism	Deicing salts, shrinkage-induced cracking	Incorporate supplementary cementitious materials and admixtures (corrosion inhibitors, SRA) to improve permeability and shrinkage characteristics
External contamination	Chloride-induced steel corrosion, sulfate attack, decalcification, and leaching	Probabilistic modeling

The concrete transport properties used in preliminary calculations should be based on those recorded for similar concrete mixtures found in the laboratory database. At the end of this phase, an interim report should be submitted with a preliminary evaluation of the service life for each mixture based on the degradation mechanisms and concrete mixture compositions.

Based on the results of Phase 1, one (or more) of the proposed mixtures per element should be selected by the design team for further testing and validation (Phase 2).

In addition, during Phase 1, it should be possible to qualify the aggregates and the binders to be used by assessing their quality and/or conformance with specific standards for the project. For example, the analysis could include an evaluation of the potential for alkali-aggregate reactivity, if this information is not readily available.

Phase 2: Mix Production and Characterization
This phase consists of the validation, optimization, and characterization of the concrete mixes selected at the end of Phase 1. It should be done in two sub-phases: one in the laboratory and the other in actual production conditions. To minimize the risk of schedule delays due to a produced mix failing to meet the project requirements, it is recommended testing more than one concrete mixture at this stage to select the most cost-effective one.

2.1 Laboratory Validation In this phase, concrete is prepared and sampled in a laboratory-controlled environment (T°,RH). The objective is to optimize the selected admixtures and constituents while meeting the requirements regarding concrete proportions. The concrete materials (i.e., binders, admixtures, fine and coarse aggregates) are usually shipped to the testing laboratory, from the selected sources are usually performed by laboratory technicians under the supervision of the concrete materials specialist. The test program should be designed to determine the main mechanical and durability properties of the selected concrete mixtures, to verify their compliance with the project specifications and evaluate the design life of elements built with these mixtures.

The program should include tests aimed at obtaining correlation data between migration test results, including resistivity tests, to establish guidelines for the quality assurance phase of the project. Also, the test program should require tests at different maturities to establish the aging factor of mixtures, which is a function of the chemical composition of the binder used and is critical for long-term service-life predictions. An example of the laboratory test program is listed in Table 11.2.

2.2 Pre-production Validation After the concrete mixtures are optimized in the laboratory and the testing program is completed, the next step in this approach requires that the batches and required test specimens be prepared by the concrete producer at the batch plant. The samples will have to be sent to qualified laboratory facility for the complete characterization of hardened concrete. Specimens should be protected adequately (all specimens individually wrapped with wet burlap and sealed in plastic bags) after proper curing. This step is important to verify that the concrete plant can produce adequately the different mixtures.

Table 11.2 Laboratory test program for physical, mechanical, and transport properties of concrete

Test	Test method	Age
Fresh properties		
Slump	ASTM C143	Upon mixing
Air content and density	ASTM C138	Upon mixing
Temperature	ASTM C1064	Upon mixing
Mechanical properties		
Compressive strength	ASTM C39	7, 28, and 56 days
Elastic modulus	ASTM C469	28 days
Tensile strength	ASTM C496	7,28, and 56 days
Drying shrinkage	ASTM C157	7, 28, 56 days. For low shrinkage, the length change should be 0.05% at 28 days
Air-void system	ASTM C457	28 days
Transport properties		
Diffusion coefficient (resist chloride ion penetration)	ASTM C1202 Or similar procedure	28 and 56 days For low diffusion, the electrical charge should range: 100–1000 coulombs
Volume of permeable voids	ASTM C1556/C642	28 and 56 days
Surface resistivity of concrete	AASHTO 358 Correlates with ASTM C1556)	7, 28, and 56 days The surface resistivity value should be >254 kohms
Freeze-thaw durability	ASTM C666	28 days The relative durability factor should be 80% after 300 cycles
Absorption and voids-moisture transport	ASTM C1792	28 days

The pre-production testing for the concrete produced at the batch plant is a simplified version of the laboratory testing program since it is performed to validate the correlation between the actual batching conditions and laboratory conditions. The test program on concrete batched at the producer's facility should be limited to fresh properties and selected mechanical properties and transport properties of concrete as listed in Table 11.2.

11.6 Calculations and Recommendations

Calculations need be performed to evaluate the durability of each tested mixture based on the time to initiate corrosion and/or the chemical degradation of the concrete. In addition to the concrete transport properties and general characteristics, the analysis should account for a variety of parameters such as the concrete composition, the concrete cover thickness, geometry of the structure or structural element, type of rebar, and environmental conditions.

Based on the calculation results, a durability report should be produced for each element of reinforced concrete structures, with test results, service life calculations, and recommendations for the final candidate mixtures to be selected for each element. At this last stage of the validation process, recommendations regarding adjustments to the mixture constituents or the design should be discussed. Report may include any of the following aspects:

1. Identify corresponding environmental exposure conditions (temperature, relative humidity) and other exposure conditions (salinity, degree of exposure). The addition of corrosion inhibitor and the determination of the required dosage.
2. Provide recommendations for more than one combination of concrete mixture.
3. Recommendations on the type of steel reinforcement.
4. Recommendations on the required concrete cover.
5. The application of a protective system.
6. Concrete fabrication, delivery, and placement procedures.
7. Curing procedures.
8. Results of probabilistic calculation showing the element will achieve 100-year service life for the reported calculation assumptions.

11.7 Quality Control During Construction

If the project documents require periodical verifications of the diffusion coefficients, based on actual concrete mixtures, this may be verified during construction. This analysis requires the determination of transport properties. For time considerations, in the context of a quality control program, only the migration and volume of permeable voids tests are performed. The maturity at which the tests would be performed would depend on the reliability of the results obtained during the qualification phase, but it will be as early as possible. Other tests, such as the resistivity, could be performed at 7 or 14 days to get an early indication of the concrete service life. As for the transport properties, the maturity at which those tests would be performed depends on the reliability of values measured during the qualification phase.

11.8 Schedule for Concrete Mixture Durability Validation Testing

The schedule for developing a concrete mixture design for design service life and durability validation may vary, depending upon the number of concrete mixtures, exposure conditions, and degrading mechanisms foreseen, for a particular project. The entire process from notice to proceed to completion of durability report is estimated to take about a year. A tentative schedule for each phase of testing is given in Table 11.3.

Table 11.3 Task descriptions and testing schedule for durability validation

Description	Schedule
Review of documents and theoretical mixture design	4–8 weeks
Mix characterization	
Analysis of project documents	
Definition of applicable exposure conditions	
Definition of degradation mechanism (durability matrix)	
Review of concrete schedule and preliminary mixture design for each element	
Mixture characterization – laboratory	
Production of a mixture in the laboratory and testing including compressive strength, modulus of elasticity, tensile strength, drying shrinkage, thermal expansion coefficient, RCPT, resistivity, freeze-thaw, diffusion coefficients	3–5 months
Analysis of results and service life calculations	
Batch plant validation	3–4 months
Production of a mixture at the batch plant and testing including compressive strength, modulus of elasticity, tensile strength, drying shrinkage, thermal expansion coefficient, RCPT, resistivity, freeze-thaw, diffusion coefficients	
Durability report	4 weeks
Total estimated duration	12 months

11.9 ACI Report on Service Life Prediction

ACI 365.1R-17 presents information on the service life prediction of new and existing concrete structures for owners and design professionals. It discusses key factors controlling the service life of concrete and methodologies for evaluating the condition of the existing concrete structures and predicting service life. It also includes the application of available methods and tools to predict the service life of new structures at the design stage. It includes examples for service life prediction and life-cycle cost analysis (LCCA) for structures exposed to degrading mechanisms (such as chloride ingress, based on current technology), and stresses the need for improvements in the modeling techniques for service lives exceeding 100 years.

The durability of concrete depends upon its constituents, cement paste, and aggregates. One of the most important factors is porosity of cement paste. The presence of excess water dilutes the paste and increases porosity, which can then be attacked by the deleterious substances and physical environment.

The durable concrete should possess the following properties:

1. A low water/cementitious material ratio, 0.38 to 0.42
2. Use minimum cementitious content of 600 lbs./cy.
3. Use SCM to resist sulfate and ASR attacks.
4. Reduce permeability of concrete by using finely divided solids materials, hydrophobic pore blockers (PRB), or crystalline admixtures.
5. Minimize drying shrinkage and associated cracking of concrete by incorporating a SRA.

6. Use concrete aggregates that are resistance to environment: abrasion and freeze-thaw weathering.
7. Use corrosion inhibiting admixtures, if corrosion of reinforcing steel is anticipated.

References

1. Dolen T Historical Development of Durable Concrete for the Bureau of Reclamation. Materials Engineering and Research Laboratory, United States Bureau of Reclamation, Denver, Colorado

Index

N. Hasan, *Durability and Sustainability of Concrete*,
https://doi.org/10.1007/978-3-030-51573-7

Printed in the United States
by Baker & Taylor Publisher Services